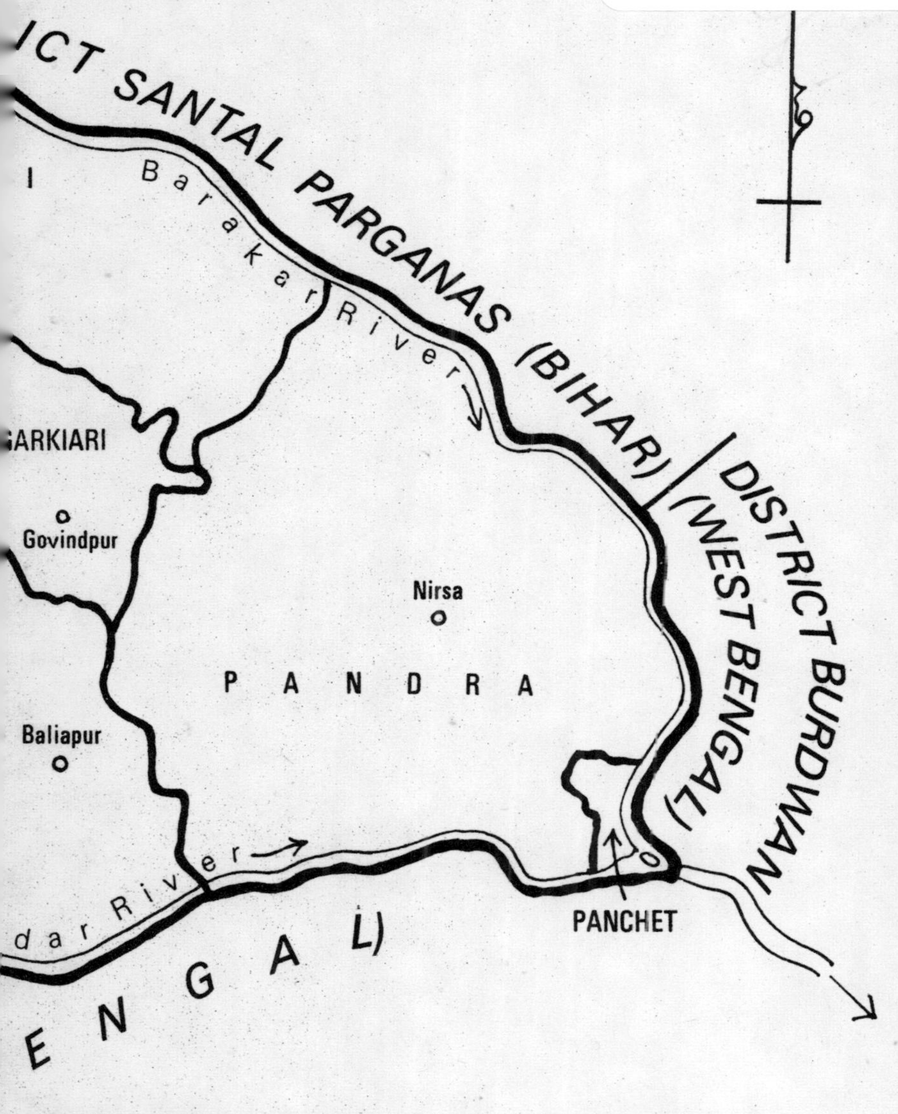

DISTRICT DHANBAD

Showing Estate Boundaries

Scale 1 inch = 5½ Miles

ZAMINDARS, MINES AND PEASANTS

South Asia Interdisciplinary Regional Research Programme
Dhanbad Research Project Report

General Editor
Dietmar Rothermund

South Asian Studies No. IX (a)

ZAMINDARS, MINES AND PEASANTS

Studies in the History of an Indian Coalfield
and its Rural Hinterland

Edited by
Dietmar Rothermund
D. C. Wadhwa

MANOHAR
1978

The South Asia Interdisciplinary Regional Research Programme of Heidelberg University (Chairman: Prof. Dietmar Rothermund) was sponsored by the German Research Council (Deutsche Forschungsgemeinschaft) and conducted in India in consultation with the Indian Council of Social Science Research.

First Published 1978

Published by
Ramesh Jain
Manohar Publications
2 Ansari Road, Darya Ganj
New Delhi-110 002

Printed at
Dhawan Printing Works
26-A Mayapuri, Phase I
New Delhi-110 064

Contents

List of Tables

Foreword

THE characteristic pattern of the modern world, industrialization, social mobility, general literacy, and new experiences for every generation originated in the Atlantic economics of Western Europe and North America. The majority of mankind in the great peasant cultures of the East was confronted with this new pattern under colonial rule or indirect forms of Western dominance. Isolated elements of the Western pattern were imposed upon the traditional societies of the East. This imposition was determined by an exogenous demand for goods and services which was reflected by a patchwork of fragmented items of Western technology and education which did not fit in with the indigenous context. Colonial rule and the partial impact of the forces of the world market stimulated a lop-sided development. The world market which was dominated by the advanced Atlantic economies drew upon the resources of the peripheral peasant economies of the East without drawing them into the main-stream of economic development. Economic growth at the periphery of the world market was not "self-sustained" and was not paralleled by the social transformation which moulded the nations of the West. As a result traditional modes of production and social structures coexisted with fragments of modernity in the countries of the East. The achievement of political independence from colonial rule or even revolutionary political change did not lead to immediate change in the social and economic situation of the great peasant cultures of the East. Their present situation is characterized by striking contrasts. Industrialization is proceeding at a rapid rate while the vast majority of the people still belong to the rural sector and are wedded to the perennial life-style of the peasantry which also influences the industrial proletariat and the urban elite.

The social scientist who wants to explain and to interpret this experience of social and economic change faces many problems. There are many theories of social and economic development, but they follow Western precedence. Moreover such theories usually have serious "re-entry" problems when they leave the rarified atmosphere of abstract concepts in order to descend to the level of the explanation of actual phenomena. This "re-entry" is difficult enough when the theory is derived from Western historical experience and applied to phenomena within the Western sphere of social and economic change. It is even more so when the theory originates in the West and is applied to phenomena observed in the East. But there is no other choice, because so far all such theories are based on Western experience. The only corrective is a close attention to historical research and empirical fieldwork which prevents the social scientist from making rash generalizations. Interdisciplinary cooperation also restrains the exuberance of the theorist, because most theories of social and economic development have a distinct flavour of the scholasticism of their respective discipline. Such scholasticism is more readily detected by representatives of other disciplines who did not have to subscribe to it in the course of their academic training.

Empirical interdisciplinary research which is aimed at identifying elements of social and economic change should concentrate on a definite region in order to provide a clear focus for the activities of various scholars. The South Asia Interdisciplinary Regional Research Programme was designed from this point of view. In this context the term "regional" was supposed to refer to an area delimited in terms of a specific research problem. The Dhanbad Research Project which was the first project under this programme, was conducted in an Indian coalfield. This site of an important primary industry surrounded by traditional peasant agriculture seemed to provide an example of the kind of problems mentioned above. The study was intended to be a contribution to fundamental research in the social sciences; it was not designed by regional planners who are asked to solve problems and to assist policy makers. The paradigmatic aspect of the study was emphasized by everybody concerned, and it is this aspect which will be stressed in this report. The project was an experiment; it did not follow a well established precedent nor was it guided by one mastermind. It was a joint venture of various scholars whose research interests and methodology differed according to their training in different disciplines. There was no leading discipline to which the others were subordinated, so as to supplement and support the trend set by that discipline. In fact, there were three major research trends which may be outlined here, because these trends have contributed to the achievements as well as to the conflicts and shortcomings of this project. These trends have also influenced the design of the project, the choice of the region, and finally the structure and style of this report. The three disciplines which set these trends were history, economics and sociology. The identification of these three trends refers to basic research interests and not necessarily to the disciplines of the scholars concerned, because the spectrum of disciplines represented in the project was much wider.

The historians were mainly interested in the impact of economic development under colonial rule on the important natural resource region to the East of Calcutta. This region was of particular interest to them because it was in many ways a most backward area into which British rule and capitalist enterprise had intruded in the nineteenth century, creating enclaves of economic activity outside the colonial metropolis. The economists, on the other hand, were concerned with the present economic structure and the relationship of urban centres with their rural hinterland. Their approach was particularly influnced by some earlier research work in agricultural economics which had established a positive impact of an industrial centre on agricultural production in its rural environment. In mutual consultations historians and economists agreed on Dhanbad District as a suitable area of research, because this district included the major Indian coalfield which had been operated for a long period of time and its rural sector would provide a testing ground for the hypothesis of the economists. The idea of concentrating on one district was mainly due to methodological considerations of field research which could not be extended too far. The historians who do not generate data in the field but have to look for them in offices and archives were less concerned about these regional limitations but agreed to focus their attention on Dhanbad, too. They shared with the economists the basic tendency to look upon this project as a regional study which should explain the specific features of economic development in this particular area. In this the historians and economists differed from the sociologists who set the third major trend in this project. They were interested in general problems such as social stratification, social mobility and the problems of equal access to education, employment, legal aid etc. They also enquired into the specific situation of traditionally underprivileged sections of the society some of whose members had achieved higher status through education and employment such as educated members of the scheduled castes or working women. These research interests were of relevance to Indian society in general, and Dhanbad was used only as a paradigm.

The project report will be presented in three volumes. This first volume will be followed by a second one with the title "Urban Growth and Rural Stagnation", edited by D. Rothermund, E. Kropp and G. Dienemann, which will contain the contributions of the economists who examined the problems of capital formation and innovation in agriculture, the availability of rural credit, and the structure of the local markets for agricultural products, but also studied wage labour in the urban sector and the types of entrepreneurs in mining and small scale industry. The third volume edited by J.P. Neelsen, deals with "Social Stratification and Political Structure" and is devoted to the theme of social dominance, discrimination, and inequality which was taken up by the sociologists and political scientists.

It is obvious that this project was not governed by any general theory of social change and economic development. In fact, field research and later evaluation was accompanied by a constant and sometimes acrimonious debate on concepts of economic dualism, dominance and dependence, external and internal colonialism etc. The dust has not yet settled and some of it will be discovered by the perspicacious reader in the volumes of this report. But considering the state of theory in this field this may not be too bad after all. Instead of pledging adherence to a single doctrine the authors have tried to bridge the gap between theoretical concepts and empirical research in various ways, subjecting themselves to the hazards of trial and error. If the reader should find that the errors predominate this would at least prevent him from making them himself.

The volumes of this report encompass only a rough cross section of the work done by the members of the research team who are listed here :

M. Ajam (Agricultural Economics)
W. Bergmann (Medicine)
G. Dienemann (Economics)
H. J. Diesfeld (Medicine)
S.N. Dutta (Medicine)
H. Haensch (Agricultural Economics)
D. Helmken (Law)
K.H. Junghans (Agricultural Economics)
D. Kantowsky (Sociology)
I. Kaul (Sociology)
E. Kropp (Agricultural Economics)
E. Kulke (Sociology)
R.N. Maharaj (Agricultural Economics)
W. Mersch (Sociology, Economic History)
J.P. Neelsen (Sociology)
H. Papendieck (Economic History)
I. Rothermund (Political Science)
G. Ruths (Economic History)
E. Moser-Schmitt (Social Anthropology)
D. Schwerin (Economic History)

G. Tiemann (Social Anthropology)
D.C. Wadhwa (Economic History)

Some of these team members have published their findings in monographs listed in the appendix of the second volume cf this report. Others were unable to complete their work because they were employed elsewhere after the three years of project finance had lapsed and could not finish writing up their reports in their spare time.

The project was planned in 1968/69 by staff members of the South Asia Institute of Heidelberg University and of the Arnold Bergstrasser Institute at Freiburg University. The first stage of field research was completed in the winter of 1969/70, and the second stage one year later. At the end of 1972 the contracts of those project members who were not on the permanent staff of these institutes expired. Most of the work on this report therefore devolved upon a few authors who could combine it with their current academic duties. This has necessarily entailed some gaps and shortcomings. But the authors concerned hope that their report will nevertheless do justice to the scope and aims of the project.

A particularly useful part of the project, the medical research on the diagnosis and prevention of filaria, a disease which is very prominent in the coalfield, is not covered by this report but the results of this research have also been published and further research along these lines is continuing at Heidelberg and Dhanbad by the respective project members.

The project could not have been conducted but for the generous financial aid provided by the German Research Council (Deutsche Forschungsgemeinschaft) under its research project programme (Sonderforschungsbereiche). The funds provided under this programme helped to finance field research in India and to employ a large research staff. Indian colleagues who participated in the project were also invited to Germany during the evaluation stage of the project. In India the Indian Council of Social Science Research was of much help in obtaining the research clearance for the project and in providing friendly support. In order to honour a promise given to the officers of that Council this report is published in India so as to make it easily available to those who would be most concerned with its findings. All German scholars who have participated in this project have benefited from the fruitful contacts with their Indian colleagues, the patience, wisdom, and hospitality of those whom they interviewed during their field research, and the cooperative attitude of all authorities concerned.

C. Tangund (Social Anthropology)

Dietmar Rothermund

Introduction

THE history of the Indian coalfield was determined by the zamindars, feudal lords who controlled land and labour and held the sub-soil rights of the region, and by the British managing agencies which extended their operations to this area. The zamindars stuck to their feudal habits, wasted their resources and encumbered their estates, the managing agencies exploited mines and men ruthlessly. The peasantry was drawn into this new economic activity at the lowest level as unskilled labour. But even the meagre wages paid for casual work in the mines were more rewarding than the hard task of converting up-land into rice terraces. Therefore agriculture stagnated and the coalfield became an enclave in a backward region. The living conditions in the hinterland of this enclave have hardly changed since the days when the mines were opened up. Even today the supply of electricity stops just a few miles outside the centre of the mining area, and bow and arrow are still in use among the local people. These are only some of the more striking examples of the contrast between the mining enclave and its rural hinterland. The present study is aimed at less obvious but even more important problems: the uses and abuses of land, capital, and labour. In order to study these problems an interdisciplinary approach was adopted which encompassed the methods of historical research as well as social science fieldwork. However, the members of the research team discovered very soon that interdisciplinary work often generates more problems than it can solve. Furthermore, there is the natural tendency that the representatives of each discipline will look upon the other disciplines as auxiliary or even ancillary partners in the quest for knowledge. The historian pursues his line of interest and asks the economist for what may be termed technical aid, the economist constructs his models and wants from the historian some data, if possible in the shape of neat time series, so as to prove his theory. If the historian strays too far into the past and tries to identify the formative stages of certain processes instead of collecting data for time series which connect the past with the present, the economist gets impatient with this useless fellow. On the other hand the historian tends to expect too much from economists and other social scientists when he hopes that they can provide readymade theoretical explanations which fit the facts with which he is concerned. Interdisciplinary work, which is started with the optimistic impression that the strength of one discipline is added to that of others soon leads to a mutual discovery of each others weaknesses. At this stage despair may overcome everybody concerned and in a pessimistic mood each participant in the interdisciplinary venture reverts to the supposedly safe guidance of his own discipline. It is only after these stages of initial euphoria and subsequent disappointment that genuine interdisciplinary work may start at a more realistic level of mutual expectations. Then there will be an increasing inclination to learn from each other and to listen to different arguments. The Dhanbad Research Project went through all these stages and it is hoped that the present report will show at least some traces of the process of mutual interdisciplinary education.

For the team of historian who contributed to this first volume of the report the process of interdisciplinary education was made easier by the fact that some of them although they were working on a historical aspect of the project had had their basic training in other disciplines such as business administration, economics, law, and sociology. But whereas the vagaries of interdisciplinary work were not a major problem for this team, the magnitude of its task and the small size of the team imposed severe limitations on its work. Furthermore, the team had not worked together before joining the project. The present author had just been appointed to the chair of history at the South Asia Institute of Heidelberg University a few month before the project was started. Detlef Schwerin was one of his first doctoral students. Henner Papendieck was a student of Professor Wolfram Fischer in Berlin at that time and was recommended to the author by him. Dr. D. C. Wadhwa of the Gokhale Institute of Politics and Economics in Poona was known to the author because of their common interest in Indian agrarian legislation. Wolf Mersch had done an M.A. in sociology and had then worked as a lecturer of German language and literature in Calcutta for some years, where he developed an interest in Indian social history, he thus joined the project in order to pursue this interest. When the team met in the field with the author who was then spending a sabbatical term at the

Gokhale Institute in the winter of 1969/70 some preliminary decisions were made about the individual research topics after an initial exploration of the source material available in the district and in Ranchi, Patna, and Calcutta. After this exploratory phase the summer term of 1971 was spent at Heidelberg. Some tentative papers were circulated for discussion and seminar sessions provided a forum for mutual information. A second field research phase was completed in the winter of 1970/71 followed by a period of evaluation at Heidelberg. In the course of this time the individual research work of the different members of the team progressed and got its definitive shape. In the following paragraphs short outlines of these individual contributions will be given so as to enable the reader to gauge the full scope of the respective work, because the present report reflects only a part of this work due to the fact that some studies have not yet been completed and others could only be summarized here.

Dr. D.C. Wadhwa found a great deal of material with the zamindars of the district and also in the Deputy Commissioner's office and in the offices of the various anchals (local administrative units). He embarked on what may be called a one man survey settlement of the larger part of the district, collected data from the records of the Board of Revenue, Bihar, and the archives of the Bengal and Bihar states, and went through the records pertaining to the numerous court cases in which the zamindars were involved. As some of these cases had come up before the Privy Council in London, he spent some time there, too. He furthermore got all his district data computerized until he could account for every piece of land in the area at the time of the survey and settlement in 1918-25 and at the time of zamindari abolition when a similar survey (bujharat) was made. In 1973, Dr. Wadhwa rejoined the Gokhale Institute to complete the remaining three volumes of his earlier work on a comprehensive survey of agrarian legislation in India. Therefore he could not evaluate all the material he collected in Dhanbad. The chapters which he has contributed to this report cover only some aspects of his more comprehensive research on which he is going to publish a series of volumes entitled "Zamindaris at Work".

Dr. Detlef Schwerin worked on the agrarian system of Chota Nagpur as a whole. His study was mainly based on government records at the divisional and provincial level, and he concentrated on the period from the middle of the nineteenth century to the first decade of the twentieth century in which the tenancy legislation for Chota Nagpur was finalised. However, his study went beyond the legal framework and was aimed at a comprehensive analysis of the relation of landlord and tenant with particular emphasis on the problems of the tribal tenantry. He completed his Ph.D. thesis on this subject at Heidelberg in 1974 and it was published by the South Asia Institute in 1977.

Dr. Henner Papendieck concentrated his attention on the British managing agencies and their role in the exploitation of the coalfield. Due to fortunate circumstances he got access to the rich archives of the Bengal Coal Company and could reconstruct the operations of some mines of the Andrew Yule Group by translating their old account books into modern accounting, a strenuous exercise for which he was well equipped due to his earlier training in business administration. He completed his Ph.D. thesis on this subject in Berlin in 1975, and it is going to be published soon.

Wolf Mersch decided to work on the history of labour with special reference to the competition for labour which prevailed in the formative years of the coalfield between the mining companies and the tea estates of Assam which had recruited labour in Chota Nagpur already long before coal mining had started. In the course of this work he got more and more interested in the early recruitment of tea estate labour and enlarged the scope of his study so as to include this aspect. During the time of field research in India he had to shoulder the additional burden of being the Field Director responsible for the financial and personnel management of the whole project. This left him little time for research work. Some of his findings were reported in an earlier cyclostyled project paper and have been utilized by other members of the team. Unfortunately he was unable to contribute to this final report.

Inspite of the fact that the team of historians was small and its research time was limited it has been able to trace the essential features of the stunted growth of the coalfield and of the hidebound agrarian system which dominated its hinterland. The present report can only provide a short summary of the findings of the different contributors, but even this summary shows that the team has concentrated on the most crucial issues.

Schwerin describes the control of land and labour in Chota Nagpur, the larger region of which Dhanbad District (alias Manbhum) is a part. The author has outlined the evolution of tenancy legislation, Wadhwa analyses land distribution under the zamindars and provides case studies of their indebtedness, and Papendieck portrays the activities of the British managing agencies in the coalfield. The main theme of persistent dualism is very well documented by these contributions. But, of course, a more comprehensive social and economic history of the region has yet to be written. Thus a study of institutions such as the office of the Chief Inspector of Mines (later on Director General of Mines Safety) or of the Indian School of Mines at Dhanbad might shed more light on the history of the mining industry. There are also undoubtedly many records of coal companies still available which have so far not been exposed to the scrutiny of scholars. The history of the district also needs further study along the lines indicated by Ranjan Kumar Gupta's recent Ph.D. thesis on the economic life of Birbhum (University of Calcutta 1976). Even the history of one zamindar's estate such as Jharia, Katras, or Tundi would be very valuable. Much has been written about the Permanent Settlement of Bengal but studies of the working of this settlement in terms of detailed accounts of individual estates are so far conspicuous by their absence. Of course, the scholar will often be put off by discouraging reports about files and accounts which have been destroyed or lost, but as the experience in Dhanbad District has shown a second or third inquiry may well lead to results which could never have been expected at the time of the first exploration. The historians' team of the Dhanbad Research Project can confidently assert that regional and local history is extremely rewarding and that there is no dearth of source material for it. If this historical part of the project report stimulates further research along these lines it will have served a useful purpose.

After this note on further historical research which will mainly interest the members of the profession we may turn once more to the question of the interdisciplinary relevance of the contributions of the historians. By focussing their attention on the formative stage of the development of the coalfield as an economic enclave and concentrating on the factors of production, land, capital, and labour, the historians have provided an adequate point of departure for the subsequent contributions of the economists which show that the pattern of dualism which was so firmly set in the colonial past to a large extent still prevails at present. Historians and economists have not written their contributions along the lines of a rigorously defined theory, because such a theory does not yet exist. Even the notion of economic dualism is very ambiguous because there are about as many definitions of it as there are theorists who have made use of this concept. Some of these definitions would be incompatible with the phenomena observed by the authors of this report. Thus the kind of dualism portrayed by Lewis who conceives of a constant supply of labour at agrarian subsistence wage levels to the industrial sector would not fit in with the Indian situation where such elements as advances and other recruitments costs or institutional wage rates for the established work force supervened and, so to speak, "distorted" the labour market. Nor would theories suffice which postulate dualism as a stage of transition in a dynamic process of economic development, a mere time lag between the opening up of a field of economic activity and the achievement of a new equilibrium at a higher level. The Indian case and similar cases of development under colonial rule require a theory of persistent economic dualism, a dualism characterized by the formation of a firm threshold which separates the relatively more advanced urban industrial sector from the rural economy, and whose visible evidence is the establishment of enclaves which greatly differ in their level of economic activity from their hinterland. This threshold, once established under colonial auspices, proves to be very difficult to remove unless the rural sector is fully integrated in an expanding national economy, which is an achievement that hardly any country of the Third World can claim even now after so many years of political independence.

In an article on "Impediments to Development from Below in India's Economic History" which I wrote when the Dhanbad Project was planned I have tried to identify some crucial constraints which contribute to the persistence of economic dualism. A short summary of the arguments advanced in this article may be useful in this context. Under British colonial rule population pressure on the land increased, a rigid revenue demand and alien legal system forced the peasants to rely on the moneylender and to depend on landlords whose hands were strengthened by the new system. Indigenous capital was "rusticated" and restricted to investment in trade or the buying of control over land. Whatever little investment in industry existed was dominated by British economic policy and the London money market. India was forced into a passive and partial relation with the world market. The credit network was operated so as to flush cash-crops out of the country, enriching the foreign trader more than the indigenous producer. Finally India was left with a rural majority of survival agriculturists. Survival rather than subsistence was their problem. The comparative comfort of adequate subsistence agriculture was a privilege of a minority of peasants who had enough land to feed their family with their own produce throughout the year. In independent India about 60% of the rural population which owned only about 25% of the land could be termed survival agriculturists. This poor rural majority which amounts to nearly half of the total population contributes probably a mere 10% to the national product, whereas about half of that product is contributed by 20% of the population who belong to the urban-industrial sector. The remaining 40% of the national product are contributed by about 30% of the population who control 75% of the land and are able to live on their own produce and to market their surplus. These striking disparities imply the persistence of economic dualism, a persistence which cannot be overcome by the forces of the market nor by mere planning or even by the investment of the relatively meagre resources of the state. Furthermore political development has proceeded along lines which prevent the emergence of developmental policies which could emancipate the rural majority, because the government depends on the productive section of the population, i.e. on the productive section as it is constituted at present due to the process which has been outlined above.

The Dhanbad Project was not planned with a view to document this process. In fact, an initial research design of the agricultural economists was based on a model of concentric circles of participation of the rural sector in the economic activity of an industrial growth centre. It was only when it was discovered that the facts did not fit this model that the project members began to reconsider their assumptions. This did not lead to the emergence of a new model and various ideas were discussed in the course of further field work and its subsequent evaluation. However, persistent economic dualism was finally accepted by most members of the project as a useful concept for the interpretation of their findings. The authors of the various chapters of this report do not always refer to this concept explicitly and their methods of analysis differ according to their academic discipline and their specific approach to the subject of their study, but there is a basic consensus about the conclusions which can be derived from this research project. The individual facets of the contributions of the various authors do not detract from the central argument, but rather help to illuminate the features of economic dualism from different angles.

1

Dietmar Rothermund

The Coalfield–An Enclave in a Backward Region

INDIA'S subcontinental shape bounded by the sea and the world's highest range of mountains suggests a uniformity which does not pass the test of closer scrutiny. India has an intricate pattern of regions determined by nature and history. The vagaries of the monsoon, great river plains and highland plateaus, invasions from the North and finally European colonial penetration from the maritime periphery have all contributed to a highly differentiated profile of this vast country. In Northeastern India regional differentiation is particularly obvious because of the contrast between the fertile alluvial lowlands, the economic base of many empires from Ashoka to Hastings and the adjacent barren uplands of Chota Nagpur, the refuge of tribal people who retained an identity separate from that of the settled peasant culture of the Gangetic plains. The rulers of the plains were usually careful not to interfere with these people as this would only have meant trouble without much benefit. But it so happened that almost all important mineral resources of India, particularly coal and iron ore, are located in this region.

Iron had been used in India for a long time. Ashoka's empire was built on a foundation of rice and iron. The art of smelting iron with the help of charcoal was known to the people of the areas where iron ore was found, and as long as iron was used only for agricultural implements and traditional weapons there was a sufficient and probably rather cheap supply of this metal available, because the simple forest dwellers who did the iron smelting bartered away their product for some food and other modest rewards.[1]

The iron they produced was soft and malleable and therefore very well suited for the purposes for which it was required in India. Therefore this indigenous industry continued until well into the nineteenth century. It was finally wiped out by British competition. But the elimination of the indigenous iron industry was not due to this competition alone. The basic problem of the Indian iron smelters was that they relied on charcoal. This fuel contributed to the soft and malleable quality of the indigenous iron but it could not provide for the hardness needed for construction, for the railways etc. Furthermore the supply of this fuel was limited. But the shift to the use of coal of which there were such rich deposits nearby was never made, as it would have meant fundamental changes in the technology of the Indian smelters.[2]

The demand for coal was therefore not generated by an Indian iron industry but by the steam engine introduced by the British for their river steam-boats and railways and for the jute mills of Calcutta. Although the coal deposits of Chota Nagpur were known much earlier, their exploitation was started in right earnest only after this new demand had gathered some momentum in the last decades of the nineteenth century.[3] Thus while the indigenous iron industry disappeared a new extractive industry developed in Chota Nagpur. But there was a vast difference between these two industries. The indigenous iron industry had provided employment for the local people, it was decentralized in many small-scale workshops and it was geared to indigenous demand. The new industry was imposed on the area, the local people were only marginally involved in it. It was concentrated in a major coalfield and it became an isolated enclave. This set the pattern for the future economic development of the Chota Nagpur plateau. The steel towns of a later period were similarly isolated. A threshold of persistent economic dualism separated such enclaves from their backward rural environment.

In contrast with urban concentrations in industrial countries such enclaves have only very tenuous links with their environment because the economic activity prevailing in them is derived from an exogenous impact. The coalfield is only one instance of this kind—there are many types of enclaves, e.g. a port serving as an entrepot for international trade, a military base maintained by a foreign power, etc. The relation of such an

[1]V.B. Singh (ed.), *The Economic History of India*, Bombay 1965, p. 27.

[2]Bridget and Raymond Allchin, *The Birth of Indian Civilization*, Harmondsworth 1968, p. 285 f.

[3]V.B. Singh (ed.), op.cit., p. 303 f.

enclave with its hinterland will be determined only by the demand for local labour and food. In exceptional cases even this relation may not exist if labour is recruited elsewhere and food is brought in from abroad. The disappearance of such an enclave would not make much of a difference to the economy of its hinterland. If the enclave does have an impact on its environment this may not necessarily be a positive one, leading only to a distortion of the economy of the hinterland without generating economic growth.

Economic enclaves which are separated by a dualistic threshold from their environment usually originate under colonial regimes or under quasi-colonial control. They are geared to a demand from abroad, e.g. mines which are operated for export only or plantations which produce cash crops for the world market. But there are also secondary or derivative enclaves which cater to the needs of the primary ones. The Indian coalfield was a typical example of a derivative enclave. Indian coal was rarely exported but used for the railways, the jute and tea industries etc. As we shall see in the chapter on mining and the managing agencies, the fact that this secondary enclave depended on the demand created in the primary ones implied that indigenous entrepreneurs could not get a foothold in this field.

Although enclaves are rarely accessible to indigenous enterprise and capital, they normally do depend on indigenous labour. But in most cases neither the skill nor the quantity of labourers in the immediate vicinity of the enclave will suffice for the work which needs to be done. Therefore labourers are often recruited at a great distance. The migration of Indian labour overseas to various plantation colonies is a case in point. The recruitment of distant or foreign labour accentuates the isolation of the enclave from its hinterland. The labourer who comes from a distance must submit to the discipline imposed on him by the terms of his contract or indenture whereas local labour is used only for casual and temporary work, its "undisciplined" character is expected and its wages are low and irregular. The people of the hinterland establish a kind of marginal co-existence with the enclave, this co-existence continuing for a long time without any change in the relationship. As K. Berrill has observed in a perceptive essay on "International Trade and the Rate of Economic Growth" enclave expansion in a low wage country which is engaged in primary productions for export does not lead to a "take-off". The home market of that country will not expand and a great deal of the income which is created will "leak abroad", as Berrill puts it.[4] This was certainly true of the Indian coalfield enclave under British colonial rule. After independence entrepreneurs from the more advanced regions of India stepped in and maintained the enclave as

[4]K. Berrill, "International Trade and the Rate of Economic Growth" in *The Economic History Review*, 2nd series, 12 (1959-60).

a going concern. Many of these entrepreneurs had been in the coal trade earlier and then had moved on to the control of production. Now the income created did no longer "leak abroad" but much of the profit made in the coalfield was invested in the more advanced economic centres of India rather than in backward Bihar, the province of the coalfield. This is what has led some observers to speak of "internal colonialism".[5] However, such phenomena can be observed also in European countries. The contrast between North and South in Italy is a famous example, where the benefits of economic growth have accrued only to the industrial centres of the North while the South remained poor and backward and its people had to migrate in order to find work elsewhere.[6] The attraction of advanced growth centres and the persistent backwardness of poor rural regions is a universal feature of economic development wherever dualism prevails. And dualism will exist as long as the respective rural regions do not experience an agricultural breakthrough.

The initiative of the state is required where the working of economic forces would merely perpetuate existing imbalances. Colonial governments did not bother about such matters, the more so if they conveniently believed in the doctrine of laissez faire and therefore held that state inference was not only useless but might even be harmful. After independence we are faced with the phenomenon of the "soft state" if we accept Myrdal's verdict in this respect.[7] But Myrdal has hit the "soft state" too hard by implying that discipline and strength of character would compensate for a meagre budget and a lack of adequate means for the enormous task of propelling millions of men across the threshold of economic dualism. Of course, state initiative has often been vitiated by corruption and inefficiency and to this extent Myrdal's criticism is justified. But even this corruption and inefficiency cannot merely be explained in terms of moral judgement. It is a part of the pattern of historical experience of the respective countries where the government has always been perched on the tip of a vast agrarian pyramid, and where the authority of government could be maintained only because it was limited to military and administrative activities in a very narrow sense of the term. Administration with a view to economic development is a new task and the scope of this task in the countries concerned would daunt even the supposedly very efficient administration of an advanced industrialized state.

[5]For a statement of the internal colonialism thesis see G. Ruths, *Regionale Entwicklung und Abhäengigkeit: Zur Theorie des internen Kolonialismus in Entwicklungsläendern, am Beispiel des Bergbau-distrikts Dhanbad/Bihar in Indien.* Unpublished Ph.D. Thesis, Heidelberg 1974.

[6]R. Jochimsen, "Dualism as Problem der wirtschaftlichen Entwicklung" in B. Fritsch (ed.), *Entwicklungsläender*, Cologne 1968.

[7]G. Myrdal, *Asian Drama—An Inquiry into the Poverty of Nations*, 3 vols., New York 1968, Vol. 1, p. 66, pp. 891-900.

The Indian coalfield and its hinterland accordingly did not show much evidence of vigorous administrative action at the time when this study was undertaken. The nationalization of the coal mines was accomplished almost immediately after the field work for this study was completed, but this measure could not be expected to bring about a sudden transformation. Another survey of the region after one or two decades would be required in order to trace the evidence for social and economic change.

After indicating the need for future research we must return to the task of reconstructing the past. In the subsequent sections of this introductory chapter we shall first try to outline the sudden growth of the enclave economy and its further development under the impact of a war time boom and inter-war stagnation and depression. We shall then turn to an analysis of the role of the great landlords who controlled the region in which the coalfield was situated and we shall close this chapter with reflections on three types of captivity—the captivity of the peasant, the captivity of the labourer, and the structure of a captive market.

The Sudden Growth of the Enclave Economy, 1902-1908

A central point of our argument about the rise of an economic enclave is the impact of an irreversible historical event which determines the pattern of economic activity for a long time to come. We shall now try to identify this event in the context of the development of the British Indian colonial economy.

In the nineteenth century British India was still relatively isolated from the world market. India's main exports consisted of agricultural produce wihch with the exception of tea was supplied by the peasantry. The peasantry was induced by the pressure of rent and revenue and the demands of the moneylender-graindealer to grow wheat and rice for the market or to cultivate cash crops such as indigo, jute, sugar-cane etc.[8] This development did not lead to the rise of enclaves, the peasant economy remained more or less unchanged except for the fact that production for the market encroached upon the subsistence of the peasantry. The storage of food-crops for the lean years was reduced and famines which had earlier been only regional phenomena became nationwide by the end of the century, a grim by-product of the growth of a national market in foodgrains.[9] The persistent depreciation of the Indian silver currency from 1876 had contributed to a secular trend of rising Rupee prices for Indian foodgrains while their gold price fell in the world market. In this way India could supply the world market with cheap grain even when there was a glut in

[8]B. Chowdhury, *Growth of Commercial Agriculture in Bengal, 1757-1900*, Calcutta 1964.

[9]B.M. Bhatia, *Famines in India, 1860-1945*, London 1963.

this market during the world agrarian crisis.[10] The extension of the railway network in India helped to establish a national market and to extract agricultural produce for the export trade even from remote areas in the interior of the country. In fact, this network was extended so swiftly in the last decades of the century that the volume of freight could not keep up with it.[11]

This opening up of India which linked the peasant economy via the railways to maritime world trade did not necessarily lead to the rise of industrial activity. Karl Marx had predicted that the introduction of the railways would lead to industrialisation because of the demand for rolling stock, machinery etc.[12] However, the British supplied all this from their factories at home. Only two items which were intimately connected with the transport and export of raw produce finally led to some rudimentary industrialisation: coal and jute. Coal was needed for the railways and in spite of the existence of large coal deposits in India, English coal had been shipped to India for a long time. Even when coal mining expanded in Eastern India, Bombay was still supplied with English coal.[13] Jute was required for the gunny bags which were used for packing grain and tea. These bags were initially supplied by the Scottish mills of Dundee, but the Scots finally decided to manufacture them in Calcutta. We are not concerned with the interesting history of the jute industry here, but we want to emphasize the relation between the growth of that industry and the rise of the mining enclave.[14] It was only when an expansion of the freight movements on the Indian railways, the concomitant demand for more coal and gunny bags, and the rise of the jute mills of Calcutta coincided that the crucial moment had come in which the fate of the Jharia coalfield was decided and a sudden spurt of production and labour recruitment set the pattern of the enclave for many years to come.

The important variables which we should observe are, therefore, the amount of freight, the number of looms in the jute industry, the production of coal and the number of workers in the Jharia coalfied. The period in which we shall observe these variables in detail is the first decade of the twentieth century. However, with regard to the development of railway

[10]D. Rothermund, "The Indian Silver Currency—An Aspect of the Monetary Policy of British Imperialism" in *Indian Economic and Social History Review*, Vol. 7, 1970, p. 91 ff.

[11]M.D. Morris and C.B. Dudley, "Selected Railway Statistics for the Indian Subcontinent, 1853-1946/47" in *Artha Vijnana* (Gokhale Institute, Poona), Vol. 17, No. 3 (Sept. 1975).

[12]K. Marx, "The Future Results of British Rule in India", in *New York Daily Tribune*, 8 Aug. 1853.

[13]V.B. Singh (ed.), op. cit., p. 312.

[14]For the expansion of the jute industry see A.K. Bagchi, *Private Investment in India, 1900-1939*, Cambridge 1972, p. 262 ff.

transport we take a longer view and look at earlier decades, too. A comparison of railway network expansion and the increase of railway freight indicates that we can identify three different trends. Before the mid-1880s the network expanded while freight movements stagnated; after that network expansion and the increase in freight more or less develop along parallel lines, but network expansion still takes the lead; from the turn of the century network expansion stagnates, but freight increases at an unprecedented scale and stimulates network expansion, thus reversing the earlier pattern.[15] This spurt in the movement of railway freight is reflected by the increase of the number of looms in the jute mills. In the first years of the twentieth century expansion is still a bit cautious from ca. 15,000 in 1900 to 18,000 in 1903, but then comes a vigorous advance to ca. 33,000 in 1910.[16] The growth of coal mining in the Jharia coalfield follows the same pattern. The field had been opened by railway extension only six years before the turn of the century. An initial spurt of production and labour recruitment in the first year had been followed by a period of stagnation (1895-1901), labour recruitment increased slightly, but production fell and reached the level of the first spurt only in 1901. Then came a second spurt in 1902 followed by three years of rather slow growth. The productivity of labour obviously increased somewhat during this period, probably due to investment in elementary machinery. The main spurt came in the years 1905-1908. Production increased steeply from 3.3 to more than 7 Mill. t and labour recruitment went up from about 40,000 to more than 70,000 miners.[17] These years were most decisive ones for the coalfield and the structure of the enclave was settled then and there. It is no accident that this period was also one of an unprecedented increase in India's foreign trade. But India's connection with the world market also meant an increasing vulnerability to economic depressions. The coincidence of an insufficient monsoon in 1907 and a world trade depression in 1908 caused a serious crisis in India.[18] The coalfield registered its impact in the subsequent year. Production declined and labour recruitment expanded only very slowly in the following two years. However, the impact of the depression did not last much longer and the expansion proceeded along the lines of the earlier trend.

A significant aspect of the formative years of the enclave economy should be highlighted here: the crossing of the threshold of local labour supply. So far we have referred to the problem of labour recruitment only in a general way. We may now try to define this threshold in quantitative

[15]M.D. Morris and C.B. Dudley, loc. cit., p. 214 ff.

[16]A.K. Bagchi, op. cit., p. 277.

[17]V.B. Singh (ed.), op. cit., p. 310 f.

[18]For the brief economic depression of 1908 see V.B. Singh (ed.), op. cit., p. 446. As far as coal mining was concerned this year was characterized by a boom and a great increase in production.

terms for the region in which the coalfield is located. In the last quarter of the nineteenth century about 20000 people left Chota Nagpur annually in search of employment elsewhere. Many of them went to the tea gardens of Assam. The demand for labour in the coalfield was not very significant in the years before the turn of the century. Furthermore, many of the earlier collieries acquired a great deal of land and tried to press the local peasantry into their service or granted land as service tenure to those who were willing to work for them. As long as mining operations did not expand rapidly this method was quite adequate. When the coal companies stepped up production and employed "raising contractors" who recruited their own labour force, the pattern changed. Contrary to what has been stated in earlier publications the local population did respond to the increasing demand for mining labour. Migration from Chota Nagpur to the tea gardens dwindled at the beginning of the twentieth century. But by 1902 the demand for labour in the coalfield had already crossed the crucial threshold. Even if the total annual stream of migrants had gone to the coalfield this would not have been sufficient any longer.[19] It so happened that the sudden spurt of 1902 coincided with a period of low rice prices which reduced the motivation of potential migrants.[20] There had been a fairly close correlation of high rice prices and increasing migration to the tea gardens for quite some time. Thus the fact that the number of miners rose to 32,000 in 1902 and that the average number of potential Chota Nagpur out-migrants of about 20,000 was probably reduced by low rice prices would indicate that this year was most important for the evolution of the enclave as far as the pattern of labour recruitment is concerned. The great expansion of 1905-1908 only reinforced this pattern and made sure that long distance recruitment from other parts of Bihar and Eastern Uttar Pradesh had come to stay.[21]

Another major feature of the enclave economy, its independence of local food supply and the concomitant lack of stimulation of local food production, was also established at that early stage.[22] Unfortunately we know very little about the state of agriculture in the area around the coalfield at the turn of the century. The first settlement report was made in 1918-25 and we can only assume that the pattern of small holdings under zamindars who did not pay much attention to agricultural improvement prevailed there throughout. A slowly growing demand for food might

[19]W. Mersch, *Dhanbad-Bericht 1972: Migration and Labour Recruitment for Coal Mines and Tea Plantations under Colonial Rule: The case of Manbhum in Chota Nagpur, India, 1881-1921.* Dhanbad Project Report, May 1972.

[20]D. Schwerin, *Von Armut zu Elend—Kolonialherrschaft und Agrarverfassung in Chota Nagpur, 1858-1908*, Wiesbaden 1977, p. 388.

[21]See the contribution of H. Papendieck to this volume.

[22]See D. Rothermund, E. Kropp and G. Dienemann (eds.), *Urban Growth and Rural Stagnation*, New Delhi 1979.

have stimulated local production, but the rapid growth of the mining enclave and the fact that this enclave owed its existence to the railways which could also bring food to this area precluded such local stimulation. We have pointed out earlier that there was a flow of foodgrains from Upper India to the port of Calcutta, and the enclave had easy access to this flow.

We may therefore assume that the local population established fairly soon a kind of marginal symbiosis with the enclave. Whosoever was in need of some additional income would put in a few days of work as temporary unskilled labourer in the enclave rather than spend his time and energy in converting upland into rice terraces in his village. Instead of stimulating agricultural production in its vicinity the enclave reduced the incentives for land reclamation or melioration by offering opportunities for casual and temporary labour. These opportunities increased with the opening up of small mines (surface mining and inclines) along the border of the main deep shaft mining belt. This casual and temporary involvement in enclave labour would support a persistent dualism. The local man would be satisfied with a wage which was only slightly higher than that for land reclamation work. The permanent miner, recruited at a long distance, would need a higher wage; often he was also given an advance which tied him to his contractor. Once the enclave was safely established in this way its demand for labour would not lead to a rise of wages in agriculture in its vicinity nor would the wages of permanent enclave labour ever approximate to the low wages of local agricultural labour.[23]

Boom, Stagnation, and Depression

After its formative years the coalfield enclave experienced three major phases of development, the war time boom of 1914-1918, post-war recession and stagnation, and the impact of the worldwide depression, 1929-1933. The war stimulated coal production because the importation of English coal became increasingly more difficult and there was a new demand for the exportation of Indian coal to South East Asia as a substitute for English coal. By 1916 coal imports were almost nil whereas exports surpassed 1 Mill. t per year. However, by the end of the war exports dropped to the same dead level as imports.[24] In fact, this wartime episode was the only one in which Indian coal played an important role as an export commodity. Otherwise the demand for Indian coal was confined to India.

Coal production exceeded 20 Mill. t. by the end of the war. Ever since 1910 increasing production had been paralleled by steadily rising coal prices.[25] Coal mining was a very good business, indeed, and the Jharia

[23]See D. Rothermund, E. Kropp and G. Dienemann (eds.), op. cit.
[24]V.B. Singh (ed.), op. cit., p. 312.
[25]Ibid.

coalfield contributed the lion's share to it. But soon after the end of the war a peculiar post-war recession hit the British Indian economy. Initially there was great optimism at the end of the war. This was reflected by an import of industrial machinery and mill work at a hitherto unprecedented scale. Pre-war imports of this type were on an average worth 50 Mill. Rs. per year. The war year had impeded such imports and therefore a higher demand was to be expected after the war. But the imports worth 200 Mill. Rs. in 1920 and more than 300 Mill. Rs. in 1921 went far beyond the level which could have been expected if entrepreneurs were only interested in catching up with pre-war standards. They obviously invested wartime profits because they were looking forward to further expansion. In this they were soon disappointed and while the machinery which they had ordered was still on its way their business declined.[26] A drop in coal production reflected this state of affairs immediately. For a short time the drop in production helped to maintain high coal prices, but soon the price of coal began to fall. Then the mine-owners stepped up production in order to maintain their income at the same level in spite of falling prices. The same strategem was later on also adopted by the jute industry and by those who grew cash crops, when the depression slashed the prices.[27] But the mine-owners resorted to this course of action about six years before the depression, and after briefly reducing coal production at the time of the depression they repeated their performance in the 1930s in the face of a further decline of the price of coal.

This development had disastrous consequences for the coalfield. Investment and conservation had never been emphasized by the mine-owners even while the going was good and rising prices ensured them of handsome profits. When the tide turned they ruthlessly exploited the coalfield. Conservation and safety were neglected. It did not matter to them whether the soil subsided, or whether subterranean fires spread for miles, as long as their slipshod methods enabled them to make some profit in spite of falling prices. The wages of labour were, of course, also affected by this change of the fate of coal mining. So far the money wages of miners had risen continuously and had topped the list of all workers in India. In the depression they lost this pride of place to the railway workers whose wages increased while the miners' money wages declined very steeply.[28] However, the miners were still better off than most other workers and the dualistic structure to which we have referred above was not affected by the depression. On the contrary, one may say that the depression confirmed the prevailing economic dualism in a most peculiar way. If only the industrial sector had been hit by the depression and the agricultural sector

[26]A.K. Bagchi, op. cit., p. 80.
[27]V.B. Singh (ed.), op.cit., 314 f.
[28]Ibid., p. 634.

had remained unscathed there should have been a reduction of the discrepancy between the two sectors. But it so happened that for no good reason Indian agriculture was even more severely affected by the depression than most industries. Agricultural prices were halved within a short period of time.[29] This could have been expected in the case of cash crops produced for export, but it applied to foodgrains which were not exported and for which the Indian demand remained as high as ever. The only explanation for this strange development seems to be that the market price of foodgrains had been maintained at an artificially high level by the moneylender-graindealer network which was linked to the world market due to the fixed exchange rate of the Indian currency maintained by the British rulers. In view of the fact that the large internal foodgrain trade of India was only marginally connected with India's export trade this was certainly a case of the tail wagging the dog. For the purposes of the present study it is sufficient to emphasize the effect of this development on the real wages of labour. Whereas real wages had declined or stagnated for several decades they now showed a positive upswing due to the slashing of the prices of foodgrains. This compensated the labourers for the decline in their money wages and maintained the dualistic threshold between industry and agriculture.[30]

The decline of agricultural prices also acted as a disincentive as far as production for the market was concerned. Only those areas which had already been geared to such production continued to supply foodgrain, because in most cases the producers were in the clutches of moneylenders and had to deliver a surplus regardless of the price which they got for it. For this reason the disastrous fall of prices did not lead to a sudden restriction of food supply which would have caused the prices to rise again. It was this constellation which made a substantial increase in the level of real wages possible, but which also prevented marginal areas from producing for the market. Thus the rural hinterland of the coalfield which was such a marginal area remained as backward as ever. As the enclave had not stimulated agricultural production in its hinterland so far it was certainly not going to do so under the conditions prevailing at the time of the depression.

The sequence of boom, stagnation, and depression had thus maintained and reinforced the pattern of isolation of the enclave from its hinterland which had been established in its formative years. By now a second generation had grown up and the coalfield enclave had become a fact of life to which everybody concerned had to adjust. Even a new wartime boom made no difference to the enclave in this respect.

[29]Dharm Narayan, *Impact of Price Movements on Areas under Selected Crops in India, 1900-1939*, Cambridge 1965.

[30]V.B. Singh (ed.), op. cit., p. 650.

War and Independence

The settled pattern of the coalfield enclave was actually an impediment to quick expansion in response to the increasing demand for coal in the Second World War. Compared to the growth of coal mining in the First World War when production increased from about 16 to more than 20 Mill. t the increase from about 25 Mill. t to 29 Mill. t in the course of the Second Word War was less dramatic.[31] In fact, stepping up wartime coal production proved to be a problem because of a scarcity of workers and the lack of machinery and because the mechanisation of the mines had not been encouraged by the preceding periods of stagnation and depression. However, the relatively slow expansion of coal mining during the war years was not only due to such bottlenecks, wartime demand was actually less than what could have been expected. India's export trade experienced an unprecedented boom during the war, but this was due to rising prices rather than to an increase in production. The jute industry, for instance, made enormous profits, the total value of its output being rated at about 480 Mill. Rs. in the first year and 570 Mill. Rs. in the last year of the war while its production actually declined.[32] This would imply that its demand for coal also declined. As we have pointed out earlier, the rapid growth of this industry in the first decade of the twentieth century contributed to the rise of the mining enclave. This important stimulus had also continued subsequently when the number of looms increased from ca. 33,000 in 1910 to about 62,000 in 1923, whereas the industry had only 68,000 looms throughout the years of the Second World War.[33] There was, however, a new customer whose demand for coal was increasing: the new Indian iron and steel industry. Indian steel production went beyond 1 Mill. t per annum for the first time during the early years of the war.[34] This meant an increasing demand for high grade coking coal and the steel producers created an enclave within the enclave by sponsoring modern deep shaft mines with a competent staff of skilled miners so as to ensure the supply of coal which suited their specifications.[35] Thus the coalfield now contained the full spectrum of mines from modern mines catering to the needs of the steel industry to small family enterprises which were operated like a treasure trove which would be emptied according to the whims of the proprietor. These small mines usually produced only soft coal used as ordinary fuel. However, coking became of increasing importance. It was done out in the open and by-products for which there

[31]V.B. Singh (ed.), op. cit., p. 316 f.
[32]Ibid., p. 271 f.
[33]Ibid., p. 272.
[34]Ibid., p. 207.
[35]Ibid., p. 313.

was not yet an adequate demand went up in smoke. The central parts of the coalfield appeared like an open-air hell with blazing fires all around.

Independence did not change the structure of the coalfield enclave overnight. The colonial pattern prevailed for quite some time. There was a slow transfer of ownership as British managing agencies withdrew from the coalfield or were taken over by Indian businessmen. But the new Indian owners simply stepped into the shoes of their predecessors. Many of the new owners had traded coal during the war and had then acquired mines as an adjunct to this business, thus in a way repeating the earlier pattern of the managing agents who controlled the demand for coal and then took over its production.

In the field of labour recruitment independent India curiously enough witnessed the implementation of a scheme which had been designed in the last years of colonial rule but which had not yet worked at that time: The Coalfield Recruiting Organization.[36] During the war the mines had experienced a serious shortage of labour due to the numerous alternative demands for unskilled construction workers etc. The new organization was supposed to ensure a continuous supply of miners particularly from the districts around Gorakhpur in Eastern Uttar Pradesh from which the coalfield enclave had traditionally got its workers. The structure of the enclave was thus reconfirmed by the establishment of an organization for the special purpose of long-distance labour recruitment. But no special efforts were made to improve agriculture in the hinterland of the coalfield so as to reduce the disparity between the enclave and its backward rural environment.

The Pattern of Settlement and Local Government

The history of the coalfield enclave is also reflected by the pattern of urban or quasi-urban settlement and the rather rudimentary forms of local government which evolved in this area. Before the rise of the enclave there were no major centres in this Northern subdivision of Manbhum District except for the modest headquarters of a subdivisional officer at Gobindpur on the Grand Trunk Road and the residences of the great zamindars such as that of the Rajas of Jharia and Katras who later on profited from the coal leases and built large palaces. When mining began the mine owners built houses for their staff and quarters for their workers on the site of the mine and thus set a chaotic pattern of quasi-urban settlement. Jharia soon emerged as a sizeable market town where traders lived who organized the food supply. The town of Dhanbad on the other hand owed its importance to the fact that it was the major railway junction on the new Grand Chord Line which connected the area with Calcutta and Upper

[36]V.B. Singh (ed.), op. cit., p. 316.

India, paralleling and replacing the Grand Trunk Road as the main artery of traffic. With the School of Mines and the office of the Chief Inspector of Mines as well as regional railway and government offices and a law court Dhanbad became an administrative and service centre quite distinct in character from the market town of Jharia or the mining settlements.

The congregation of a large number of people in the rather limited area of the coalfield enclave should have called for some organization of local government which could have taken care of the infrastructure which was needed for the support of an urban population. The British had introduced municipal government in India and, therefore, the required organizational framework was readily available.[37] But the mineowners were averse to municipal taxes and preferred their quasi-urban chaos to any form of local government. Finally a Water Board and a Board of Health were established in order to provide for the most elementary needs of the growing population.[38] Only Dhanbad with its urban service population attained the status of a municipality, but there was not much scope for taxation in this town anyhow because it had hardly any industry and not much trade. In striking contrast with Tata's steel town Jamshedpur some hundred miles further south which was a neat township controlled by the steel company, the coalfield enclave was in a mess before as well as after independence because nobody could hope to get anything done against the wishes of the mine-owners who thought that all they needed to do was to provide some elementary housing for their workers.

The miners' rows with which the whole area was studded at irregular intervals were, indeed, solid and spacious when compared with the local mud huts. They were built of stone and had a high vaulted ceiling, but the main problem was that they were usually overcrowded, serving as dormitories for several men instead of as a house for one family. Most miners kept their families anyhow in their distant villages and were satisfied with this dormitory style of living while they worked in the coalfield. They never made the coalfield their home and looked forward to the time when they could return to their villages. Similarly the engineers and managers who resided in large bungalows in the mining compounds looked upon their years in the coalfield only as a transitory phase of their life. Nobody had roots in the enclave except for the coalfield rajas whose splendid palaces looked somewhat incongruous in this environment. When their fortunes declined and these palaces decayed they blended more harmoniously with the general tone of grimy misery which prevailed in the coalfield. As

[37]For an account of the early evolution of municipal institutions in British India see D.E. Wacha, *Rise and Growth of Bombay Municipal Government*, Madras 1913.

[38]The Jharia Water Board and the Jharia Board of Health are quasi-municipal bodies. The chairman is usually a local politician who enjoys the power and patronage provided by this office. The jurisdiction of these boards covers only part of Dhanbad District. Funds are provided by the mining industry.

superior landlords who controlled the whole area including the subsoil rights the rajas were in a position to dictate their terms and to impose a better pattern of settlement and local government on the area under their jurisdiction. But they missed this chance and signed away their rights for a song in the formative years of the enclave. Their successors who might have known better were not able to recover lost ground, the coalfield engulfed them and they remained there like insects preserved in a piece of amber.

The Zamindars of the "Jungle Mahals"

In order to understand the behaviour of the coalfield rajas we must refer to their historical background. The details of the agrarian system of Chota Nagpur will be explained in the subsequent chapter. Here we shall only deal with the background of those zamindars whose estates cover the coal belt and whose ancestors had been the lords of the land long before coal mining was started in this area. These lords of the land were proud of their Rajput origin. The Mogul regime had recognized them as zamindars and the British continued this tradition. The Permanent Settlement of 1793 also covered their estates but they had additional privileges which other zamindars of Bengal did not have. Their estates were recognized as quasi-territorial units and were therefore deemed to be impartible. Thus they escaped the fate of sub division and fragmentation which quickly overcame many estates of the Bengali zamindars.[39]

The special privileges bestowed on the rajas of what were termed the "Jungle Mahals", were, of course, in a way similar to the privileges of lions in a zoo; they were protected but their freedom was curtailed. The British saw to it that they were even protected from the consequences of their own improvidence. Except for a brief period in the beginning of the nineteenth century their estates remained impartible and as such were earlier treated as inalienable, because it occurred to British judges only very much later that impartibility did not necessarily imply that the estate concerned could not be alienated.

This position of the zamindars of the "Jungle Mahals" was in striking contrast even with that of their immediate neighbours. The Muslim zamindar of Rajnagar, who controlled a large part of Birbhum District from which the Jungle Mahals had been separated only a few years earlier, provided a remarkable example of what could happen to a local warlord under the new dispensation. After the severe famine of 1770 he had tried hard to meet the British revenue demand although his peasantry was decimated and much land had remained fallow. Finally the peasants led by their headmen rose in rebellion against him. His estate was sold up for

[39]For a study of these estates see contributions of D.C. Wadhwa to this volume.

arrears of revenue and subdivided in many parts which were bought by "new men". Some of these were his former agents while some were servants of the government who used their position to get hold of some land. By 1800 the old estate was completely dismantled and the former zamindar replaced by 233 new ones.[40]

But in the "Jungle Mahals" the old zamindars retained their estates which were often encumbered and then put under government management. However, these estates were not operated as single units; they were soon affected by the universal phenomenon of sub-infeudation. The zamindars created many tenure-holders and gave land grants to Brahmins and others who were in their favour. These tenure-holders were really the effective landlords whereas the great zamindars were overlords with not much of direct contact with the peasantry. A tenure-holder as later on defined by Bengal tenancy legislation was a man who held 30 acres or more from a superior landlord and who was therefore deemed to have tenants himself. This definition was introduced so as to provide for the acquisition of occupancy rights by these tenants, a complicated problem with which we are not concerned here.[41] But the category of tenure-holder is of importance in the context of this chapter because it encompasses the middle strata of rent receivers who dominated the countryside. In the impartible zamindaris of the "Jungle Mahals" they practically played a similar role as the "new men" did who divided the Birbhum estate among themselves. There was also another common feature: ordinary zamindaris of Bengal as well as subordinate estates such as those of tenure-holders were liable to frequent subdivision not merely by sale but due to the pattern of inheritance. The effective landlord was often one of a group of quarrelling brothers and cousins or an agent of a clan of absentees. When zamindari rights were abolished in Bihar after independence the tenure-holders were also eliminated on payment of compensation and while there were only a few great zamindars in Dhanbad district (*alias* Northern Manbhum *alias* Jungle Mahals), there were several thousands of tenure-holders who had to be taken care of. In some places more than seven strata of sub-infeudation could be observed in this district.[42] All these rent receivers were satisfied with a share of the surplus of the peasantry. Subsistence agriculture so to speak also included the subsistence of rent receivers who did not act as improving or capitalist landlords. Even elementary common assets such as village tanks, embankments or other minor irrigation schemes were rarely maintained properly,

[40]R.K. Gupta, *The Economic Life of a Bengal District: Birbhum 1793-1857*, unpublished Ph. D. Thesis, University of Calcutta 1976.

[41]For a discussion of the definition of the "tenure-holder" see D. Rothermund, "The Bengal Tenancy Act of 1885 and its Influence on Legislation in Other Provinces" in *Bengal Past and Present*, No. 162, 1967.

[42]See the contributions of D.C. Wadhwa to this volume.

because under this agrarian system the truly effective landlord was the exception rather than the rule.

When coal mining started in the district the coal companies inserted themselves into this structure at the intermediate level of tenure-holders with leases obtained from the great zamindars. A British administrator has reported the story of one of those rajas who signed numerous leases without knowing what he was doing as his corrupt agent told him that the gentlemen concerned had heard of his beautiful handwriting and were eager to have a specimen of his signature.[43] To the zamindars these new tenure-holders were very welcome because they provided them with a larger rental income; for the peasantry these new landlords did not make much of a difference initially as they were not interested in agriculture. Unlike other earlier coalfields where mining companies acquired landed rights in order to convert peasant holdings into service tenures for miners, the quick expansion of mining operations in the Jharia field necessitated a different approach to the recruitment of labour as we have described above. Therefore, the peasantry was treated with indifference by the mining companies and villages were disturbed only when they blocked the way for railway sidings, pitheads etc. or when they were later on affected by the subsiding of the soil when empty inclines collapsed. Thus the coalfield enclave included quite a few villages whose straw thatched huts and green rice terraces provided a striking contrast to the grimy mines next door.

In a way the new extractive industry blended with the prevailing agrarian system as a kind of subsoil agriculture, and just as much as the zamindars did not interfere with their agrarian tenure-holders as long as they paid their rent they did not mind what their new tenure-holders were doing. Nor did the zamindars make any attempt to try their hand at coal-mining. Their Rajput life style was anyhow not suited to business ventures, and as the subsequent analysis of the managing agency system will show, they were probably well advised not to start coal-mining on their own. However, if the Rajput rajas had been replaced at an earlier stage by "new men" who were closely related to trading interests the pattern might have been different. Such people would have kept a closer watch over the new ventures in their charge.

The Rajput zamindars of the coalfield and its immediate vicinity also did not respond to the new demand for agricultural produce created by the growing labour force of the mining enclave. They might have resumed land which had been subjected to sub-infeudation and cultivated it with hired labour, growing rice and vegetables for the expanding local food market. But they did nothing of this sort and the pattern of sub-infeudation was not

[43]F. Bradley-Birt, *Chota Nagpore—A Little Known Province of the Empire*, London 1903.

affected at all as a detailed survey has shown. There was no trace of a capitalist or entrepreneurial approach to agriculture in spite of the great demand for agricultural produce which was there for everbody to see who had a look at the market of Jharia.[44]

In summing up we may state that the crucial events which determined the social and economic profile of this region were the creation of the Jungle Mahals as a reservation for traditional Rajput feudalism protected by the institution of impartible estates at the beginning of the nineteenth century and the sudden growth of the coalfield enclave in this peculiar setting in the first decade of the twentieth century. The contrast between these two worlds was not marked by a tragic conflict of values or a violent clash of interests but rather by an odd co-existence of palace and pithead, ricefield and coalfield, the local peasant and the migrant miner.

The Determinants of the Enclave Economy: Three Types of Captivity

In this introductory chapter we have tried to define the main features of the enclave economy, described its particular manifestation in the Jharia coalfield and illuminated the historical background of the local zamindars who controlled the area concerned. The subsequent chapters will be devoted to some essential determinants of the enclave economy. There is first of all the agrarian system of Chota Nagpur which regulated the control of land and labour in the hinterland of the coalfield. The analysis of this system will show that the prevailing feudal structure precluded an adequate response to the challenge of the new economic activity and placed severe constraints on the development of the region. The second determinant which will be discussed in detail is the British managing agency system which was geared to a restricted market and to the making of huge profits with a minimum of capital investment.

Both systems, the feudal agrarian one as well as the colonial manifestation of capitalism in the managing agency system, were of a narrow and restrictive type and mutually reinforced each other in maintaining the enclave character of the coalfield and the backwardness of its rural hinterland. They did not permit genuine economic growth and stifled regional development.

The third determinant which must be mentioned here is the colonial system of labour recruitment which was at first shaped by the demand for Indian labour abroad and in the tea plantations of Assam and then made its impact on the long distance recruitment of labour for the coal-mining enclave. All three determinants are characterized by the common feature of captivity as contrasted with freedom. The agrarian system is based on

[44]See the chapter "Zamindars and their Land" by D.C. Wadhwa.

the captivity of the peasant who owes his lord rent and service. The managing agency system operates in a captive market where supply and demand do not respond to each other freely but must be manipulated so as to match each other. The labour recruitment system depends on the captive worker, who is bound by contract, indenture, and advances and thus depiived of his free mobility.[45] This type of captivity was, of course, more obvious in the recruitment of plantation labour whereas the recruitment of miners seemed to be less oppressive. But even in this case most labourers were not really free agents and masters of their own fate. They fled from rural poverty and captivity only in order to submit to a new servitude under tough "raising contractors" who made their profit by exploiting their labour force as much as possible. Max Weber had postulated that untouchables and members of the depressed classes of rural society would look for an escape from rural servitude and join the industrial labour force.[46] Morris D. Morris has controverted this thesis in his study of textile workers in Bombay by showing that members of landowning peasant castes rather than untouchables were working in Bombay.[47] An analysis of the composition of mining labour, however, would prove Weber's thesis. The arguments of Weber and Morris can be reconciled with each other by stating that the better type of industrial labour market of the textile mills and similar urban factories was open to those people who had some rural resources to fall back on and therefore could choose their job whereas the untouchables, the landless, and the rural poor had no choice and had to accept the meanest job wherever they could find it, or, to put it in our terms, they could only make the transition from one captivity to another.

The three types of captivity which have been mentioned here form an interdependent pattern, they were sub-systems of the colonial system . In the following chapters they will be analysed separately. In a concluding section we shall take up the theme of this introduction once more and make an attempt to link the historical evidence with the findings of the economists who have surveyed the contemporary scene.[48]

[45]H. Tinker, *A New System of Slavery—The Export of Indian Labour Overseas, 1830-1920*, London 1974.

[46]M. Weber, *Gesammelte Aufsäetze zur Religionssoziologie*, Vol. II: *Hinduismus und Buddhismus*. Tüebingen 1963 (3rd edit.), p. 11 ff.

[47]M.D. Morris, *The Emergence of an Industrial Labour Force in India*, Berkeley 1965.

[48]For the contributions of the economists see D. Rothermund, E. Kropp, and G. Dienemann (eds.), op. cit.

2

Detlef Schwerin

The Control of Land and Labour in Chota Nagpur, 1858-1908*†

Introduction

The Chief Minister of Bihar while inaugurating a minor irrigation scheme in Hazaribagh recently said that the region "though it was very rich in mineral resources and had a number of major and big industries, remained backward owing to non-availability of irrigational facilities".[1] Government investment in heavy industry in this region totals Rs. 1200 crores but by all accounts the economic condition of the majority of its

*The following discussion appears in a more detailed and more extensively doumented form in my book *'Von Armut zu Elend, Kolonialherrschaft und Agrarverfassung in Chota Nagpur, 1858-1908'*. Wiesbaden 1977.

†The following abbreviations have been used in the footnotes.

AC Assistant Commissioner; BR Board of Revenue; CCN Commissioner of Chota Nagpur; DC Deputy Commissioner; GD General Department; GOB Government of Bengal; GOI Government of India; JD Judicial Department; L Lohardugga; LD Legislative Department; LGOB Lieutenant Governor of Bengal; R Ranchi; RAD Revenue and Agriculture Department; RD Revenue Department; SAC Senior Assistant Commissioner; SBR Secretary of Board of Revenue; SC Special Commissioner; SCROB State Central Records Office Bihar; SGOB Secretary to the Government of Bengal; SOCN Settlement Officer of Chota Nagpur; SR Settlement Report.

[1]*The Statesman Weekly*, 21 Feb. 1976.

people, who still work in agriculture, remains stagnant.[2]

Two questions come immediately to one's mind: 1. Why do these sizeable investments in industry have no appreciable spill-over effect on the living standard of the rural population, and 2. Why is irrigation in Chota Nagpur still conspicuous by its absence? I do not intend to answer the first question which refers to a phenomenon we are only too familiar with in many developing countries. To answer the second question one has to look back over time to understand the man-made obstacles to the spread of irrigation and to the intensive cultivation of land. Although the majority of Chota Nagpur's people have always depended for their livelihood on the produce of land which will not bear an ear of grain without water, major parts of the region are without any irrigation facilities. These wholly rain-fed areas are at the mercy of the monsoon. The result, and this can be traced back as far as the 19th century, has been insufficient grain production, indebtedness, systems of forced and bonded labour and a constant drain on the region's man-power through emigration.

Grain production, which by the beginning of this century did not meet even the most modest food requirements of the rural population in about half of the region of Chota Nagpur, was constrained by an oppressive land tenure system, the consequence of Hindu pressure on an ancient tribal society. Hindu migration into Chota Nagpur over the last three hundred years (17th-20th centuries) transformed land holding from a communal ownership system to the quasi-feudal hierarchical pattern of the Gangetic plains. This new order which during the 19th century could count on the support of the colonial power, syphoned off the meagre surplus of the region. It was squandered in consumption by a minority which not only did not invest in land, but found it to its advantage to stop the peasant from trying to improve the land and his own situation. Due to the way in which the system worked, the growth of population was not matched by an otherwise possible expansion of agricultural production, thus resulting in a downward path from poverty to misery. To follow this development in some detail is still of relevance because it was accompanied by bitter antagonism between immigrating Hindus and the Scheduled Castes and Tribes. As far as can be judged from the outside, this sharp cleavage persists and has not lost its economic foundation.[3] It is a feature one must reckon with in Chota Nagpur.

The Region and its People

In the Indian context Chota Nagpur seems to be insignificant as far as

[2]The Statesman Weekly, 9 June 1973.

[3]Vide a description of bonded labour in Palamau: *National Labour Institute Bulletin*, Vol. 2, No. 10 (Oct. 1976), pp. 387.

size and population are concerned. However, at 27615 sq. miles (the size of Ireland) it is bigger than 38 member-states of the United Nations. Its pre-1956 boundaries included the districts of Hazaribagh (NE), Manbhum (E), Palamau (NW), Ranchi (SW), Singhbhum (SE) and the two small principalities of Kharsawan and Saraikela. Even forty years ago (1931) its population of 6.83 million was larger than those of seventy-eight independent nations of today. However, its population density has always been quite low. Although Chota Nagpur's population increased during the period 1881 to 1931 by 57.2% which was far above the average of Bihar and Orissa (+16.8%), its population density in 1931 was at 246 people per sq. mile, well below the 379 people/sq. mile of the province as a whole. It can be assumed that the comparatively high growth rate of the population was caused by the low density. This fact was even more marked in the neighbouring princely states of Orissa, where the density was much lower than in Chota Nagpur, but where the population nearly doubled between 1881 and 1931. Chota Nagpur's population growth of an annual mean of 1.2% over the fifty-year period, modest as it may seem by today's standard, was achieved against a background of significant and constant population losses. From 1891 to 1931 outbound migration was twice as large as inbound migration. For 1921 this meant for instance that 750,000 people who had been born in Chota Nagpur, and most of whom were of tribal origin, lived outside the region, approximately half of them in the tea districts of Assam and Rajshahi, whereas only 324,000 people, born across the borders, had moved into Chota Nagpur. But outbound migration was much more significant at an earlier stage, its growth rate diminished over the years, whereas inbound migration grew much faster.[4] This indicates increased employment opportunities within the region due to new jobs in coal-mining (major expansion since 1894) and in the Jamshedpur steel industry (since 1907).

Chota Nagpur is a rugged country which till the 1860's was covered by dense forests, so much so that the Muslim rulers of northern India knew the region by the name of "Jharkhand" (forest country); similarly the British called part of it "Jungle Mahals". Even during the first quarter of this century 44% of the region was still covered by jungle.[5] The forest spread mainly across the north and south of Chota Nagpur. Part of it was of very poor quality, other areas had however been preserved by protective legislation since the 1870's. Jungle, hills and non-navigable rivers were the factors responsible for the rather limited communication between Chota Nagpur and the outside world. It improved to some extent only subsequent to the advent of the railways in 1871. In the following fifty years, especially since 1889/91, Chota Nagpur got a basic rail grid of

[4]Schwerin (1977), pp. 51, 371.
[5]Ibid., p. 25.

approximately 1300 km in length, connecting the region with the rest of India. However, neither the still rather limited network of British-Indian roads and railways nor British law affected the indigenous tribal society so profoundly as the continuous though slow migration of Hindus from the Gangetic plains into Chota Nagpur.

The indigenous tribes of Chota Nagpur are remnants of a pre-Aryan population which survived in this mountainous fastness by being shut off from the cultural, economic and political mainstream of India. Thirty-three different groups of non-Aryan origin could be identified in 1931, half of which were in a more or less advanced stage of Sanscritization. The groups range from the Savars, a tribal group of only a couple of hundred in number, to the Santals and Oraons, both of whom numbered more than half a million members in Chota Nagpur alone. Many of these groups have their own pantheon and speak languages that are completely different from the surrounding major North Indian languages like Hindi, Bengali and Oriya. However, all of them shared a common culture which made them appear very different from their Hindu neighbours. In 1901, they comprised more than half (57.2%) of the population of Chota Nagpur. Their share declined slightly over the next few decades.

As is to be expected, the Hindus were largely concentrated in those parts of Chota Nagpur which adjoined Hindu majority districts. Chota Nagpur can be roughly divided into three zones which cut across the region. In the north the Hindus formed the majority (up to 90%) of the population. In the middle strip between north and south they averaged between 30% and 50%. In the large southern part of the country they were only between 10% and 30% of the population. However, the percentage breakup does not relate directly to the influence the Hindus wielded nor to the power they enjoyed. It merely tells something about the migration movements into a certain subregion. But I should like to propose the following hypothesis: (1) At one point tribal groups formed the majority of Chota Nagpur's population everywhere. This is, inter alia, born out by the fact that the petty rulers of pre-British days, who later turned into estate owners under the Permanent Settlement of 1793, were nearly all of tribal descent. This was true even for an area like Northeast Hazaribagh (Kharagdiha) where in 1901 Hindus made up 80% to 90% of the population, but where the landed class was of Bhuiya extraction. (2) The Hindus migrated into this once completely tribal and only sparsely populated country over a long period of time partly pushed by the shrinking economic opportunities of the plains, partly pulled by offers of tribal chiefs who slowly abandoned their own gods and cultural identity and turned towards the gods of the Hindu pantheon, Shiva, Kali and Jagannath. The number of Hindus in a given subregion was determined by its proximity to the Gangetic plain and by the length of time this subregion had been exposed to immigration. Regions located more to the South were exposed to Hindu migration at a later stage than the North of Chota Nagpur which bordered on Hindu majority districts. Thus, whereas the full impact of Hindu immigration on tribal society in Ranchi and Singhbhum was felt only after British rule had been established, the Hindus had already succeeded in superimposing their own social and economic structure over the Northern districts of Palamau, Hazaribagh and Manbhum well before the grant of the Diwani (1765).

The Economic Situation of Rural Chota Nagpur

The agrarian structure of Chota Nagpur—as, indeed, all over India—is characterised by the small size of holdings. In 1961, between 62% (Ranchi) and 83% (Dhanbad) of the holdings were not larger than 5 acres. Astonishingly enough a very similar pattern existed in the Palamau Government Estate of 1897 and the Palamau District in 1961. In both cases 15% of the holdings were below one acre and 47% to 49% between one and five acres in size. Though the larger holdings had been more numerous in 1897, the overall distribution of land was very much the same. Unfortunately, only one other similar comparison can be made which extends to the Ranchi Sadr Subdivision. The Settlement Officer, J.D. Sifton, surveyed 240 villages in 1911 and found that roughly 40% of the farms were larger than 8 acres contrasting sharply with only some 15% in 1961. The decrease in the size of holdings over the last fifty years due to increasing population pressure was to be expected and is surely to be regarded as the normal case. Palamau seems to be the exception. Sifton had calculated that an adult needed at least 700 g cereals per day taking the minimum requirements as established by the Famine Code and the Prison Administration as his guidelines. On this basis an average rural family of 5.4 members in the Ranchi Sadr Subdivision needed 14 acres of land to produce enough food and marketable surplus to survive in the most modest manner possible; this calculation takes into account the prevailing soil quality and the necessary expenses for rent, clothes, salt, oil etc. According to Sifton's survey of 240 villages in six out of ten thanas of the Ranchi Sadr Subdivision only 6.5% to 20% of the holdings amounted to 14 acres or more and were thus sufficiently large to support an average sized family. Sifton came to the conclusion "there is no reason to suppose that the conditions of other thanas are much better. It follows from this that the majority of the cultivating families of the district cannot make a living from agriculture".[6] Twenty years before, another officer B.C. Basu had arrived at similar conclusions. He compared grain demand and supply on a family by family basis in several villages west of the area covered by Sifton. His final verdict was that "the great

[6]*Ranchi Settlement Papers:* 5.10.1911, J.D. Sifton SOCN, Note on Subsistence Holdings in Ranchi.

TABLE 2.1
SELECTED ECONOMIC DATA, CHOTA NAGPUR 1900-1938

Subdivisions, their location within district, period of survey and settlement operations	*Non-urban population/sq. mile net area sown*	*Culturable waste as share of arable and potentially arable land (1)*	*Husked rice/day and head of non-urban population*	*Don (low-land)/head of non-urban population*	*Bakasht as share of total arable land (Don Bakasht as share of total Don)*	*Landlords, share of total agri-cultural product (2)*	*Irrigated area as share of total arable land*
		%	*g*	*acres*	%	%	%
(1)	(2)	(3)	(4)	(5)	(6)	(7)	(8)
Gumla (NW-Ranchi) 1927-35	433	22.0	655	0.62	11.6	.	0.1
Gumla (NW-Ranchi) 1902-10	442	21.1	628	0.60	14.7 (17.6)	19.2	.(3)
Simdega (SW-Ranchi) 1927-35	448	35.4	646	0.61	10.5 (.)	.	0
Simdega (SW-Ranchi) 1902-10	468	38.3	562	0.53	16.4 (21.2)	18.6	.(3)
Khunti (S-Ranchi) 1927-35	498	20.4	570	0.61	10.5 (.)	.	—
Kolhan (a) (S-Singhbhum) 1913-18	525	23.0	1012	0.76	—	3.5	6.4

Contd. . .

TABLE 2.1 *Contd.*

(1)	(2)	(3)	(4)	(5)	(6)	(7)	(8)
Latehar (S-Palamau) 1913-20	534	47.1	214	0.32	10.7 (11.3)	20.8	2.4
Porahat (a) (W-Singhbhum) 1900-03	556	51.1	.	0.51	1.4 (. .)	.	2.6
Ranchi Sadr 1927-35	563	19.7	651	0.64	10.2 (.)	.	0.4
Kharsawan (b) (N-Singhbhum) 1925-27	580	20.5	.	0.76	5.2 (.)	.	2.3
Khunti (S-Ranchi) 1902-10	592	27.8	560	0.54	11.3 (14.0)	14.9	.(3)
Saraikela (b) (N-Singhbhum) 1904-07	594	34.7	.	0.82	5.3 (.)	.	0.5
Kharsawan (b) (N-Singhbhum) 1904-07	612	36.6	.	0.70	6.3 (.)	.	1.1
Ranchi Sadr 1902-10	632	21.8	597	0.60	10.7 (11.6)	17.5	.(3)
Saraikela (b) (N-Singhbhum) 1925-28	633	23.5		0.77	5.2 (.)	.	7.5

Contd . . .

TABLE 2:1 *Contd.*

(1)	(2)	(3)	(4)	(5)	(6)	(7)	(8)
Barahabhum/ Patkum (S-Manbhum) 1907-12 (a)	687	26.0	.	0.59	7.5 (7.7)	12.1	1.6
Dhalbhum (E-Singhbhum) 1934-38	707	31.0	.	0.68	5.8 (.)	.	1.3
Dhalbhum (E-Singhbhum) 1906-11	760	46.2	.	0.66	5.3 (6.1)	.	0.7
Palamau Sadr (N-Palamau) 1913-20	782	37.9	164	0.23	27.9 (34.5)	46.9	13.4
Manbhum Sadr 1918-25 (c)	790	27.8	554	0·61	16.7 (17.4)	23.3	20.5
Hazaribagh Sadr 1908-15	810	47.3	381	0.43	14 1 (14.2)	22.9	.(3)
Giridih (E-Hazaribagh) 1908-15	914	43.6	345	0.37	13.4 (13.8)	22.5	.(3)
Chatra (W-Hazaribagh) 1908-15	978	58.8	271	0.37	22.4 (23.5)	41.4	.(3)
Dhanbad (N-Manbhum) 1918-25	1458	35.8	286	0.31	10.8 (11.3)	17.8	11.0

(a) Estate (b) Princely State (c) Manbhum Sadr Subdivision excluding Barahabhum and Patkum Estates.

(1) arable and potentially arable land refers to land which is either already under cultivation or in the opinion of the settlement officers could in future be used for cultivation (= culturable waste).

(2) excluding lac, tusser and straw.

(3) the irrigated area comprised 0.1% of arable land in Ranchi (1910) and 1.1% in Hazaribagh (1915).

The figures presented in columns 2 to 8 are based on Schwerin (1977) table 11 (= col. 3), table 12 (= col. 8), tables 13,58, appendix E (= col. 2), table 29 (= col. 7), table 32 (= col. 6), table 58 (= col. 5), table 62 (= col. 4).

body of raiyats cannot indulge in what will be called a luxury in this country of a full diet of cereals all-round the year." The population substituted its grain requirements by collecting forest products like Mahua flowers and wild yams. Of necessity the people were vegetarians most of the time.[7] Again some 60 years earlier S.T. Cuthbert commented in 1827 on the living conditions in Ranchi: "The peasantry, generally speaking, do not appear to enjoy a state of great comfort, their huts are miserable and their ordinary food is of the poorest kind."[8] It is quite obvious that all through the 19th century material conditions of the raiyats in Ranchi were rather critical even to the point that the majority of the population was not able to produce enough food for its annual consumption needs. In a way, this is surprising because Ranchi has to be considered a grain surplus district due to its export trade in grain.

Land in Chota Nagpur was broadly classified as low-land (Don, 41% of arable area) and up-land (Tanr, 59%).[9] Low-land contains normally superior soil, is levelled, embanked and is exclusively used to grow rice. Up-land is less fertile, does not keep moisture so well and is used for the cultivation of an inferior quality of rice and of millets and pulses. Low-land, like up-land, is further subdivided each into three classes. Large parts of up-land (Tanr III) which comprise approximately 30% of the total arable area[10] cannot be cultivated on a yearly basis but have periodically to lie fallow to recover. It is, however, possible to convert up-land into third-class low-land (Don III) by levelling and embanking it. This enables the land to keep the monsoon precipitation for a longer time and thus to grow high yielding varieties of rice.

Chota Nagpur subsists on what is called rainfed agriculture. In the last four decades of British rule on an average only 5% of arable land was irrigated.[11] Whereas more or less no irrigation existed in Hazaribagh and Ranchi, a fair amount of land was irrigated in Manbhum (15%) and Palamau (10.4%). In Singhbhum again the area irrigated (4.1%) remained insignificant (vide table 2.1, col. 8). Tanr like Don III was normally not irrigated and had to be content with what the monsoon had in store. The average paddy crop of low-land in Chota Nagpur was estimated on the basis of crop-cutting results of the three classes of low-land (Don I-III) which were conducted on an intensive scale during the settlement operations in the first three decades of this century. To see whether the overall foodgrain requirements of the population were matched by sufficient production, the paddy crop was converted into husked rice and finally shown as rice-availability per capita per day. The resulting figures which exist for all

[7]B.C. Basu, *Report on the Agriculture in the District of Lohardaga*, Calcutta 1891.

[8]S.T. Cuthbert, "Report about the People and the Lands of Chota Nagpur" in: *Journal of the Royal Asiatic Society*, Vol. 8, 1846, pp. 407-416.

[9]Schwerin (1977), p. 37.

[10]Ibid.

[11]Ibid.

sub-regions except parts of Singhbhum (vide table 2.1, col. 4), were compared with the grain requirements per person as used by J.D. Sifton. The settlement officer had estimated that 175g of the 700g grain needed per day per person would be produced by up-land; that leaves 525g of grain that have to come from low-land. It is assumed that up-land is sufficiently productive to provide for 175g grain or pulses per day per person although this is not quite so obvious because the per capita endowment with up-land in Hazaribagh, Manbhum and the greater part of Singhbhum was lower than in the Ranchi Sadr Subdivision which provided the basis for Sifton's calculations. However, the main question is whether rice production reached the level of 525g per person per day. It turned out that the southern part of Chota Nagpur, Ranchi, South Manbhum and probably the whole of Singhbhum produced enough for its local rice requirements. But Palamau, Hazaribagh, and the Dhanbad Subdivision of Manbhum on account of its mining population were faced with a substantial rice deficit during the first quarter of this century. This picture is partly verified by statistics of railborn grain transports. Although figures exist only for a limited number of years (1894-1899, 1912-1922) it is quite clear that the inflow of grain into Chota Nagpur by rail was much greater than the outflow.[12]

This means that the South of Chota Nagpur, comprising 58% of the region's population (1921), produced enough grain for its own requirements and for some export trade, while the North did not. Even in places like Ranchi, where overall grain production was sufficient, the vast majority of rural households suffered from a food deficit because of the limited size of holdings. The poor soil conditions of Chota Nagpur required larger than average size holdings to sustain peasant families at the most modest level of self-sufficiency throughout the year. It can reasonably be suspected that this picture of an agrarian economy, which did not meet the basic requirements of the majority of the people, is not only valid for the early 20th century, (the period of the settlement operations provided for the first time the necessary economic data), but also reflects the actual living conditions during the greater part of the 19th century.

Indebtedness and Bonded Labour

In keeping with this situation the majority of Chota Nagpur's peasants (raiyats) were indebted. However, the data on this problem are extremely scarce. Normally there exist only general descriptions speaking of "nearly every raiyat" or "seven out of eight farmers" needing new loans on a year to year basis to carry on farming. The manager of the Dhalbhum Estate

[12]Schwerin (1977), p. 360; *General Administrative Report for the Division of Chota Nagpur 1864/95-1898/99*; Government of Bihar and Orissa, *Report on the Trade carried by Rail and River in Bihar and Orissa*, Patna 1922.

(Singhbhum) wrote at the time of the Dufferin Enquiry of 1888 that according to a current proverb the raiyats of the estate tilled their land only for the benefit of the mahajan (moneylender).[13] The situation was not quite as straightforward, because a few years earlier a district officer had held exactly the opposite view citing the lack of emigration as proof for his theory.[14] Observers agreed on the situation of the Kolhan Government Estate (Singhbhum) where indebtedness was reported to be either completely missing or minimal. The picture is less debatable in Manbhum. In 1888 a sampling of 550 artisan and peasant families showed an indebtedness of 65%; of the 277 peasant families an even higher percentage of 75% was indebted.[15] The higher share of indebted raiyats is a general feature of the Indian agrarian economy as has been shown by the All-India Rural Credit Survey of the Reserve Bank of India in 1951/52. The greater security provided by the possibility of land mortgage is probably the explanation for the higher percentage of indebted raiyat families. Enquiries conducted in 1929 by the Bihar and Orissa Provincial Banking Enquiry Committee and after independence by the Reserve Bank of India establish an indebtedness of well above 70% of the sample families in Chota Nagpur. The situation in this region might have been a little worse than elsewhere in India, but it was in no way very distinct. Debts were more frequently incurred for unproductive than for productive purposes; interest rates, already outrageously high, were even more demanding than in neighbouring regions. Interest accounts were fraudulently manipulated to the disadvantage of the debtor. Large debts were usually secured by mortgage with possession which in Chota Nagpur was practised in two ways. In the form of *Bhugut Bandha* capital and interest were paid out of the produce of the mortgaged land. The land reverted to the owner after a time period previously agreed upon. In the case of *Zarpeshgi* the produce of the land paid only for the interest. The principal had to be paid back separately before the land reverted to the owner. In practice, of course, the creditor took permanent possession of the land, and this was probably understood from the very beginning.

The Zarpeshgi form of mortgage with possession was so much to the advantage of the creditor and his position so strong that the approximate 18:1[16] preponderance of Zarpeshgi over Bhugut Bandha was only to be expected. Having said this, I must add that the percentage share of mortgaged

[13]GOI, RAD (Famine) Dec. 1888: No. 351 R, 31.5.1888, C.C. Stevens, CCN to SGOB, RD.

[14]*Report on Wards' and Attached Estates in the Lower Provinces*, 1883/84, p. 92.

[15]GOB, JD, 402, 16-18, May 1875: 31.3.1875, Note by E.T. Dalton; GOI, RAD (Famine) Dec. 1888: No. 351 R, 31.5.1888, C.C. Stevens, CCN to SGOB, RD; Craven (1898), para. 82.

[16]Average of data pertaining to Ranchi (1910, 1935), Dhalbhum (1911), Barahabhum and Patkum (1912). Schwerin (1977), p. 327.

land to the total of raiyati low-land[17] was not very impressive. In Singhbhum (1911) and Palamau (1920) mortgaged land made up less than 1% of raiyati low-land; in West-Ranchi (1910) it reached slightly more than 1%; only in Manbhum (1925), Khunti (1910) and the Ranchi Sadr Subdivision (1910) did it average between 3% and 7%.[18] In Palamau and Chatra (Hazaribagh) the percentage share of mortgaged land to raiyati low-land was extremely small reflecting, according to the settlement officers, the low collateral value of land caused by insecurity of tenure and high land rents.[19]

However, these two regions of Northwest Chota Nagpur were plagued by another even more onerous form of indebtedness, namely bonded labour or *Kamiauti*. Against small advances of an average 17 Rupees (Palamau 1918) the debtor pledged his labour against the interest due until he was in a position to repay his debts. But "once a kamia, always a kamia", as the settlement officer, T.W. Bridge, remarked.[20] Kamiauti was an old institution which already existed in the later district of Ranchi, when S.T. Cuthbert wrote his much quoted report on Chota Nagpur in 1827.[21] In the first quarter of this century it prevailed mainly in Northern Palamau and North Chatra[22] where roughly 10% and 25% respectively (the whole Bhuiya population of North Chatra) of the people had sold themselves into or had inherited kamia status. The malik, the creditor, could transfer his title which amounted to a virtual sale of the raiyat and depressed the kamias to a slave-like status. Furthermore the creditor's title was enforced by the police and the local courts to prevent a breach of "contract". This, of course, ran contrary to the East India Company's Slavery Abolition Act of 1843 (Act V of 1843, Sec. 2).[23] The kamia's living condition was miserable in the extreme. He received a minimal wage in kind, a dole, as Bridge called it, in case his malik had work for him. As he was not allowed to leave his village his opportunities to find additional wage labour were in the most favourable case severely limited. The kamia and his wife could count

[17]This statement presumes that most of the mortgaged land was low-land. If the mortgaged land is calculated as a percentage share of total raiyati land (low- plus upland) the share would diminish significantly, e.g. in the Ranchi Sadr Subdivision (1910) from 5.3% to 2.1%.

[18]No comparable figures are available for Hazaribagh (1915). It was said, however, that the situation was not too different from Ranchi or Manbhum, with the exception of the Chatra Subdivision. Sifton (1917), p. 103.

[19]J.D. Sifton, *Hazaribagh SR*, Patna 1917, p. 104; T.W. Bridge, *Palamau SR*, Patna 1921, p. 117.

[20]*Palamau Settlement Papers*: No. 1025, 28.8.1918, T.W. Bridge, Settlement Officer Palamau to P.W. Murphy, Director of Land Records and Surveys, Bihar and Orissa; Bridge (1921), p. 94.

[21]Cuthbert (1846).

[22]Sifton (1917), p. 111.

[23]*Palamau Settlement Papers*: No. 7 P, 1918, P.W. Murphy, Director Land Records and Surveys to SBR, Bihar and Orissa.

[24]Vide note 20.

on 360 work days between them. In addition, they cultivated one-sixth of an acre of up-land which the creditor had put at their disposal. Altogether the average 4.5 person kamia household was able to earn 41% of its annual grain requirements as defined by the "Famine Code". The resulting very substantial food deficit was balanced as far as possibly by roots and fruits collected in the jungle.[24] This degrading system of labour—the kamias were called the "hands and feet" of the malik, the "breath of his nostrils"[25] —was widespread in North Palamau and Chatra and reveals something about the economic situation of these subregions.[26]

Before leaving the question of indebtedness, it is interesting to note that contrary to general belief professional moneylenders played only a minor role as creditors of the Chota Nagpur peasantry. In Manbhum (1925) for example 90% of the creditors were labelled "peasant" and only 5% as "mahajan". Similarly only a small share of the mortgaged land, in Ranchi 7.1% (1935) and in Dhalbhum 12.9% (1938), had gone to professional moneylenders.[27] It has been said that "the rise of a moneylender class among the tribals was the result of the restrictions imposed on the transfer of holdings from tribal to non-tribals"[28] under the protective agrarian legislation of the early 20th century. I am convinced that those acts changed very little as far as the creditor community of Chota Nagpur was concerned. I base myself *inter alia* on the statistics of the "Registration offices" which show for the five years immediately preceding the first bill which tried to channel land-transfers that even at that time professional moneylenders were in a minority of 5%-30% all over Chota Nagpur as buyers of raiyati land.[29] This and the fact that the mahajans were in any case over-represented because of their preference for registered land transactions proves to my mind conclusively that raiyats or tenure-holders themselves were the main creditors of their neighbours. Professional moneylenders in Chota Nagpur were not so much interested in peasant land as in the socially more attractive position of a landlord.

Patterns of Migration: The Search for Work

At the beginning of this century the economic situation in Chota Nagpur

[25]*Hazaribagh Settlement Papers*: 26.5.1941, A Note on the Kamiauti System of Thana Partabpur by P. Kapur, Assistant Settlement Officer.

[26]Unfortunately neither the general economic situation nor the life of the agricultural labourer seem to have greatly improved over the last half a century in Palamau. *Kamiauti* is still widespread. See the heartbreaking accounts of R.N. Maharaj and K.G. Iyer in: *National Labour Institute Bulletin*, Vol. 2, No. 10 (Oct. 1976), p. 387.

[27]Schwerin (1977), p. 320.

[28]S.K. Singh (ed.), *Tribal Situation in India*, Simla 1972.

[29]Schwerin (1977), p. 323

was thus characterized by insufficient overall grain production in half the region, holdings which did not produce enough to meet the food or other requirements of the peasant family and a widespread and inescapable indebtedness which took the vicious form of *kamiauti* slave labour for some unfortunate 60,000 people in Palamau (1918) alone. In order to survive people were therefore forced not only to tap jungle sources for food, but to seek wage employment. Not too many alternatives existed for the willing manual labourer (coolie) within the region. He could try to find employment as a field hand who was however mainly in demand during the agricultural peak season. His daily monetary wage, if he was lucky enough to find work at all, was however one of the lowest in India around the turn of the century.[30] Moreover, his real wage, when measured in rice, had been declining since the 1870s.[31] Another substantial employment opportunity within the region opened up only after 1894 with the dramatic expansion of coal-mining in Dhanbad. This immediately affected outbound migration from Manbhum which increased from 1891 to 1901 by less than the Chota Nagpur average and declined in absolute numbers in the next two decades. The inbound migration into Manbhum increased and included a growing number of workers from neighbouring Hazaribagh.[32] In earlier times when employment potential within Chota Nagpur was still very limited people had to look for work in the indigo plantations of Bihar, the jute mills and the docks of Calcutta, and, later on, in the tea-gardens of Assam. This search for work, especially during the agricultural slack season, was commented on as early as 1827 by S.T. Cuthbert. He reported that the adivasi workers from Ranchi were in great demand because they worked for less wages and harder than others. He added that half of an average family of four to five stayed at home to look after the fields, the rest went in search of work.[33] We may assume that the seasonal character of migration changed after 1850 subsequent to the rise of the tea-industry which offered work contracts of longer duration (3-5 years). The tea gardens of the Duars and of the valleys of Assam and Surma became the main and continuing attraction for Chota Nagpur labourers in search of employment although wages in mining were since 1897 distinctively higher than in the distant tea industry.[34]

The rise and decline of migratory movements closely reflected the results of harvests. Climbing rice prices indicating scarcity pushed the coolies

[30]Schwerin (1977), p. 289.

[31]Ibid., p. 292. Dharma Kumar, *Land and Caste in South India*, Cambridge, 1965, was able to show a similar decline in real wages for South India. Kumar (1965) p. 166. However, wages in Chota Nagpur's agriculture were customarily paid in kind. Although less data on grain wages exist the trend of decreasing wages in kind is distinct, too. Schwerin (1977), p. 296.

[32]Ibid., pp. 51, 118, 121, 371.

[33]Cuthbert (1846).

[34]Schwerin (1977), p. 383.

out of their villages and farmsteads and into the nets of "cooly recruiters" who whisked them away to the tea gardens.[35] The labour demand of the tea industry was so great that a brisk and lucrative recruiting trade developed. The "Horo Akirinko", the men-seller as the Mundas called the recruiter, bagged Rs. 50 to Rs. 120 for a contracted coolie at the turn of the century.[36] This was a substantial premium and led directly to shameful excesses. The government responded by passing eight regulatory and preventive enactments during the period 1863 to 1901, but could claim only partial success.

Actually, the people of Chota Nagpur should have responded in a different way to their economic plight. After all, Chota Nagpur was, compared to neighbouring regions, only sparsely populated. Instead of migrating to other areas they should have improved their land by irrigating it or converting up-land into third class low-land (Don III) thus raising unit productivity. Another alternative was land reclamation from jungle or wasteland. The potential of these three alternatives was vast in most cases. One must ask why these opportunities for capital formation through additional labour were lost.

Although generally speaking Chota Nagpur was only thinly populated, the density of its non-urban population per square mile net area sown was comparatively high (vide table 2.1, col. 2) In the first quarter of this century population pressure on "net area sown" reached in Chota Nagpur with exception of West Ranchi already 50% and more of the average density per square mile net area sown of India in 1970 (=1004 people/sq. mile).[37] In fact three quarters of Chota Nagpur's population lived in 1921 in those parts of the division where the pressure on land (net area sown) was 60% and more of the average Indian density of 1970. It is not so much the fact of this comparatively high density that is astonishing, but that at the same time large tracts of land waited to be taken under the plough. The case of the Chatra Subdivision in West Hazaribagh, a desperately poor area, is the most puzzling of all. With a high density of 978 people per square mile of net area sown (1911), 59% of the land which could have been used for agriculture, was allowed to lie idle being thinly covered by unproductive and deteriorated forest that made up 86% of the culturable waste. In other areas the discrepancy was not quite as obvious, but in all cases one-fifth of the land marked by the settlement officers as being arable or potentially arable land remained unploughed and unsown (vide table 2.1, col. 3). 67.8% of the culturable waste was covered by rather meagre forest. It cannot be denied that some of this land

[35]Schwerin (1977), p. 389.

[36]J. Hoffmann, et al., *Encyclopedia Mundarica*, 13 vols., Patna 1950, p. 156; SCROB: No. CR 457, 13.4.1899, H.C. Streatfield, DC, R to CCN; SCROB: No. 618 J, 4.12.1899, A. Forbes, CCN to SGOB, G.D.

[37]United Nations, *Statistical Yearbook for Asia and the Pacific, 1974*, p. 126, 130.

was probably of marginal character. However, one should note that there was a definite trend in the pattern of land reclamation in the course of time. In all cases where data for different periods of this century could be compared, density per square mile of net area sown decreased,[38] although population increased substantially over time. To illustrate this point let us consider the case of Ranchi, the largest district in Chota Nagpur. There the pressure on land declined from 540 people per square mile of net area sown (1910) to 490 people (1935); rice production consequently rose from 581g (1910) to 631g (1935) per head and day, although the non-urban population grew by 11.6% during the same time. This of course indicates that the net area sown expanded faster (+22.8%) than population and that most of this increase was in the category of the more productive low-land which went up by 20.9% whereas up-land gained only 0.6%.

At the same time the ratio of culturable waste to arable and potentially arable land decreased only from 26.2% to 23.7%, which indicates that cultivation became more intensive rather than extensive. These few figures prove sufficiently that under the right circumstances the people of Chota Nagpur were eager to increase agricultural production by converting up-land into low-land (Don III) and by clearing wasteland where this proved to be rewarding.

If land reclamation proceeded so vigorously in later times why was it not taken up at an earlier stage? What prevented the peasantry of Chota Nagpur from increasing production, an urgent necessity for the majority of these raiyats? Why did such great numbers have to leave their land throughout the 19th century, instead of improving and expanding their holdings? In short what impeded the raiyats from taking advantage of the existing economic opportunities within Chota Nagpur itself? To answer these questions we must take a close look at the socio-economic development of the region and the resulting institutional impediments to agricultural production.

The Socio-Economic Framework of Agrarian Production in Chota Nagpur

I have pointed out earlier that I attribute great importance to the fact that in Chota Nagpur the social structure of the tribal population was affected by the rising tide of Hindu immigration. It can be shown that the process of Hindu penetration started with varying intensity not later than the early 17th century, superimposing Hindu concepts of society and stratification upon a generally egalitarian tribal society.

[38]This is true for Ranchi (1910/1935), Kharsawan (1907/1927), Dhalbhum (1911/1938) and others, but not for Saraikela (1907/1928). The case of Saraikela is special because of its proximity to the Tata Steelplant in Jamshedpur and the increased inflow of migration. Otherwise its situation would not have differed from the rest of the region (vide table 2.1, col. 2.).

Tribal Organisation and Feudal Subversion

The base of tribal society, the village community, was made up of the descendents in the male line of the village founder, the khuntkattidars, and the relatives on the female side, the parjas.[39] Some indispensable craftsmen of semi-Hindu origin like potters, smiths and weavers rounded off this tightly knit community. The khuntkattidars were the joint owners of all village land whether cultivated, waste or jungle. They were allowed to cultivate as much land as possible. A periodic redistribution of arable land did not normally take place.[40] The clan (kili) of the khuntkattidars as well as the whole village were led by a religious (pahanr) and an originally less important secular head (munda). These leadership functions had become hereditary within the families (khunt) of the eldest and second eldest sons of the village founder. Important decisions affecting the whole community were, however, taken by the village assembly (panchayat) which was customarily presided over by the pahanr and where all men of the village enjoyed equal deliberating and voting rights. The Pati or Parha transcended the narrow confinement of the village power structure in grouping ten to twelve villages together. In ancient times the pati probably served defence purposes. One of the village pahanrs or mundas headed the pati with the title of manki. The basis for adhering to a certain pati was the fact, at least originally, that all the khuntkattidars of the village group belonged to the same kili. Out of this comparatively democratic power structure with oligarchic overtones emerged a centralised authority over several patis and sometimes vast stretches of territory in the shape of local tribal rajas. This institution probably grew out of the manki system. At one point the most powerful manki assumed the role of a "king" over his less powerful fellow mankis. Very soon the new raja took on the trappings of Hindu royalty and the door was flung open to Hindus from the plains who offered services and merchandise. Out of necessity and custom they were rewarded with the grants of villages which originally entitled them to no more than the collection of traditional tribute (chanda). Chanda was not an assessed tax, but a "voluntary" contribution of the villagers to the expenses of central authority. The levy was passed on from the raiyat to the village headman, then on to the manki and finally went into the storehouse or treasury of the raja. Somewhere in this chain the new link of the Hindu tenure-holder or Diku (foreigner) was forged. The more middlemen between the peasant and the raja were added—in Chota Nagpur the maximum number was five, which compared favourably with other

[39]In the following brief account of tribal society a terminology originating with the Mundas was used. Other tribal groups have evolved similar institutions but named them differently.

[40]Hoffmann (1950), p. 2389.

Indian regions—the more burdensome the system became. Each one in the chain, though not so much the traditional leaders like manki and village headmen, was living from the one directly below him in this hierarchy which in the end meant an increasing burden for the raiyat. This phenomenon, of course, was well known all over the subcontinent and was terminated only with the land reform measures of post-independence India. For Chota Nagpur, however, the gradual establishment of the middlemen/tenure-holder hierarchy introduced a disruptive element in tribal society and revolutionized the whole relationship between the people and the land. When by the middle of the 18th century the British arrived in the region, tribal society had already been transformed to a large extent in Northern and parts of Eastern Chota Nagpur so as to conform with the rigid and exploitative pattern of the Hindu landholding system. The process had gone so far that the tribals had even lost their former numerical superiority. We are concerned here with the effects of these changes on the land system and consequently on agricultural productivity. One of the things the non-tribal Diku middlemen had to do in order to increase his hold over his tenure was to eliminate the competing power of the traditional village and circle headmen. Once this goal of driving out or reducing the influence of the most powerful and well-to-do men in the village was achieved, the remaining weaker sections of the population became a comparatively much easier prey for the Diku tenure-holder. In the village or villages under his direct sway he could then try to convert chanda into a regularly assessed land rent, to raise traditional labour dues (Beth Begari), to resort to other forms of levies (Abwabs) and to get hold of the more productive land for his own cultivation (Bakasht). Finally he would try to oust the remaining recalcitrant tribal raiyats and replace them with more co-operative Hindu settlers. In this process he would normally be backed by the Hinduised petty ruler of the area. Later on he could count on the ignorance of the British courts. Litigation became a powerful new weapon in his hands. His illiterate raiyats turned into helpless victims of the courts which in turn were manipulated by him.

One of the decisive actions of the colonial power which influenced the region's land system until Independence was the extension of the Permanent Settlement of 1793 to Chota Nagpur. The apex of political power, the petty rajas, by a stroke of the pen became estate proprietors, owners of the soil in a sense which had never occurred before or ever applied to them. Similarly their subjects became dependent tenants, who were only allowed usufructuary rights to the land. The idea of communal landownership resting with the khuntkattidars was completely foreign to the British concept of landlord and tenant, which was to prevail for the next one and a half centuries. One can readily understand how the interests of the Diku middleman and the concept of British landholding were detrimental to tribal society. If tradititional tribal society had not already disintegrated before the advent of the British, as in the case of Northern and Eastern Chota Nagpur it was bound to do so under the combined impact of tenure-holder and British law.

The Landlord's Exactions

Let us now take a closer look at the way in which the Diku tenure-holder who was the effective landlord, tried to appropriate his tenants' labour. This is important, because I believe it had a profound impact on the region's agricultural productivity. First we must consider the question of rent, whether in kind or cash, which the raiyat was obliged to pay to the landlord. Similar to rent payments was the levying of abwabs or *rakumats*, which were also paid in cash or kind. However, rent was in some way connected with the amount of land the raiyat cultivated whereas abwabs were in most cases assessed per household. We may then turn to the inter-dependent problem of forced labour and the landlord's constant encroachment on raiyati land which he expropriated and turned into his own bakaṣht land. I believe that the ever growing bakasht area led to rising demand for forced labour or, as it was called in Chota Nagpur, for beth begari. The combination of compulsory labour without remuneration and bakasht land which could thus be cultivated at no or very little cost to the landlord provided the most important part of the landlord's income. The landlord of Chota Nagpur acknowledged this fact in saying that half of his income resulted from rent, the other half from free labour rendered to him, or as he put it "Adha dam, adha kam".[41]

Figures collected during the land settlement operations of the early 20th century proved the overriding importance of the bakasht crop for the landlord's income. At that time the income of landlords at the bottom of the tenure-holder hierarchy was roughly divided between the following main sources: up to 5% of his income was derived from produce rent (other than from bakasht sharecroppers), 12-41% from money rent, and 55-88% from bakasht land.[42] It is necessary to point out that only the first two to three layers in the tenure-holder hierarchy had to rely on this resource pattern. The higher one moved, the less important the income from bakasht land became. On the other hand, different new sources of income opened up such as forest dues, royalty from mining rights etc. Even then, for most of the estate proprietors at the top of Chota Nagpur's social pyramid, money rent receipts, either from their tenure-holders or directly from their *khas raiyats*, remained the single biggest item of income till the "Zamindari Abolition Act" of post-independence.[43] To put it briefly, for

[41]GOB, RD, Aug. 1870, 64-67: No. 3, 30.4.1870, A.D. Haldar to H.L. Oliphant, DC, L; GOB, JD, Apr. 1859, 77-79; 25.3.1859, CCN to A.R. Young SGOB; SCROB: No. 167 OR, 12.3.1890, E.G. Lillingston, DC, L to CCN.

[42]Schwerin (1977), p. 104.

[43]Ibid., p. 216.

the small landlords and therefore for the majority, the combination of bakasht land and beth begari was of prime importance, for the bigger landlords rent was the most important factor which determined their income.

Rent

In 19th century Chota Nagpur the most visible source of socio-economic friction was the persistent drive of the landlords to convert the insignificant chanda payments into land rents where this had not been done before. These rents, in theory calculated per unit of land and according to soil quality, were not automatically adjusted to changes in the cost of living, which rose all over India especially towards the end of the century. Therefore any enhancement of rent, under the conditions of Chota Nagpur, became at once a power contest between the landlord and his tenants. Although legal procedures for rent increases had been established in 1859, British courts were of ambiguous importance in the rent struggle.

Originally, of course, all rent payments were remitted in kind. Produce rents in most cases were either a fixed amount of grain (*saika* system) or half the crop (*baku* system). Already in the first quarter of the last century, however, rent payments were generally made in cash.[44] One hundred years later in most areas produce rents were payable for not more than 2% of total arable land. This share increased to roughly 3%, if we consider only low land under produce rents. The exceptions, as in many other cases, were Chatra (NW Hazaribagh) and North Palamau where produce rents had to be paid for 4% and 15% respectively of the low land.[45] We know that in the second half of the 19th century the peasantry, generally speaking, favoured cash rents. Although most of the rents were already paid in cash, several cases are known where either produce or mixed cash/grain rents were recalculated as cash rents. The first years of this century, on the other hand, saw frequent attempts by the landlords to reintroduce produce rents because they were unaffected by inflation and since they were significantly higher than the prevalent cash rents. This is especially true for North Palamau, where the landlords tried to create a *fait accompli* by introducing produce rents shortly before the land settlement operations started in 1913.[46]

I have mentioned above that in theory cash rents were calculated per unit of land and soil quality. In reality, however, rents were the result of a compromise between the landlord and his tenant and reflected rather the

[44]Cuthbert (1846); A. Prinsep, "On the Trace's of Feudalism in India . . ." in *Journal of the Royal Asiatic Society*, Vol. 8, 1846, p. 390-406.

[45]Schwerin (1977), p. 160.

[46]Bridge (1921), Appendix, p. 65; *Palamau Settlement Papers*: 21.3.1917, T.W. Bridge, Produce Rent in Pargana Japla; *Hazaribagh Settlement Papers*: 20.2.1908, Note by K.C. Ghosh, Manager Encumbered Estates Hazaribagh.

degree of pressure the landlord could exert over his tenant than any other criteria. One of the reasons why it was impossible to arrive at just and equitable rates of rent was the absence of any standardised land measures. The traditional measures like powa, kat, anna and others normally defined an area by the amount of seed grain required to sow it. In practice, the extent of the unit of land thus described would greatly vary even within the same village. Although equal rates of rent for instance per powa might have been applied, the question of the extent of the area covered by the powa remained unresolved under this system and was left to "bargaining" between landlord and tenant. Normally, the amount of rent was determined by the kind of the coercion which the landlord could bring to bear upon his tenant but he increasingly tried also to exploit the British legal system for his own purposes. This was especially true for Ranchi and Manbhum, where during the 1880s and 1890s the average number of rent suits per year increased by 117% and 79% respectively. These two districts experienced a significantly higher level of rent litigation than the rest of Chota Nagpur. This is clearly reflected by the following average numbers of rent suits per annum:

	Ranchi	*Palamau*	*Hazaribagh*	*Singhbhum*	*Manbhum*
1881-1884	713	49	693	19	1832
1885-1889	1100	45	890	47	2450
1893-1897	1276	134	915	173	2535
1898-1902	1550	159	918	310	3275

Less litigation, however, cannot be interpreted as an indication of a reduced degree of tension, but must simply be seen as an expression of the comparatively more powerful position of the landlords in Palamau and Hazaribagh,[47] which made their recourse to the courts less necessary.[48] It is difficult to gauge the impact of the legal machinery on ordinary life. British courts were partial, which stemmed from the complexity of legal procedures, ignorance of the people and their customs and the prohibitive cost of litigation. Although the interference of courts in rent disputes showed very marked regional imbalances, litigation as has been said, was a powerful new weapon in the hands of landlords, who were in 99 out of 100 cases the plaintiffs. It worked even without going to court. The mere threat to sue was often intimidation enough. If that proved insufficient the landlord would use the court decrees, which were generally in his favour, and were obtained in one test case to threaten the rest of his tenants.

[47]The situation in Singhbum was different due to early land settlement operations and government controlled rent collection.

[48]*Hazaribagh Settlement Papers*: 4.4.1914, Extract from the Director Land Records Inspection Note.

To sum up, in 19th century Chota Nagpur rent was mainly paid in cash and its amount was normally based on nothing other than the tenant's limited capability to pay. The century was marked by a continuous struggle to increase rent. Since 1859 rent enhancement without the mediation of the Revenue Court had become illegal. Although most disputes between landlord and tenant continued to be decided by force outside the British legal system, the courts became increasingly involved, especially in Ranchi and Manbhum. In practice British jurisdiction turned out to have a definite anti-tenant edge.

Perhaps even more oppressive than the system of rent assessment was the levying of abwabs or, as these dues were also called, *rakumats*. The abwabs were extortions in addition to the rent and normally paid in kind. They were levied in irregular intervals and under different pretexts. A landlord might have asked for a goat, a maund of paddy, a *paila* of ghee or oil as a contribution to his wedding, the birth of his child or to an annual religious festival like Dasahra or Durga Puja. It has been said that it was especially objectionable for adivasis to contribute to the homage of Durga and Kali whom they did not worship themselves.[49] Although most of the abwabs seem to have been levied in kind, there were also extortions in cash. Government cesses, which the landlord was obliged to pay, were passed on to the tenants; if the landlord purchased an elephant, he created the *hatti chaul*; if it was to be a horse, he established the *ghora dana*,[50] so if he granted a rent receipt, he asked for the *Nagdi Abwab* (as in Manbhum where it was levied at the rate of 6% to 12% of the annual rental).[51] Twenty-four different varieties of abwabs have been counted and the list is far from complete.[52] The British took a strong stand against abwabs by a ruling in 1793 which said that existing levies had to be amalgamated with the rent and that the imposition of new abwabs was forbidden.[53] However, this rule scarcely affected the relation of landlord and tenant. During the following one hundred years, the colonial government and its local representatives vaccilated between paper commitments against abwabs and a strong tendency to go along with anything that could claim to be based on custom, especially if custom was pleaded by the economically stronger section of the population. It has been pointed out that during the 19th century the demand for abwabs in Ranchi grew constantly.[54] Most of these obnoxious

[49]SCROB: 24.9.1890, F. Hahn, G.E.L. Missionary to S. Bailey, LGOB.

[50]E. Lister, *Hazaribagh Gazetteer*, Patna 1917, p. 121.

[51]B.K. Gokhale, *Manbhum SR*, Patna 1928, p. 66.

[52]Bhattacharji, *Guide and Glossary to the Survey and Settlement Operations in the Chota Nagpur Division*, Patna 1939, p. 21.

[53]*Regulation VIII of 1793*, Sec. 54, 55.
Regulation IV of 1794, Sec. 2.
Regulation V of 1812, Sec. 2.

[54]J. Reid, *Ranchi SR*, Calcutta 1912, p. 82.

levies disappeared with the land settlement operation of the early 20th century, but its real end came only with the "Bihar Land Reform Act" after independence.

Beth Begari

As already mentioned earlier, the Chota Nagpur landlord used to say that half of his income was generated by compulsory labour rendered to him by his tenants without remuneration. Forced labour or Beth Begari was considered to be a standard obligation of the tenant, besides the payment of rent and abwabs. It was normally expected during the agricultural peak season, thus diverting the peasant's labour from his own fields to the bakasht land of his landlord at the most critical time of the year. Traditionally only 14 working days[55] were due from the tenant who tried, mostly in vain, to resist any demand beyond this norm which was sanctioned by custom. The landlords on the other hand always attempted to live up to their convenient expectations according to which only their own needs as defined by themselves could effectively determine their demand.[56] It is probable that factors like the overall balance of power between landlord and tenant, the demand situation on the landlord's bakasht land and the proximity of the tenant to the landlord's land were more relevant in determining the amount of compulsory labour services than any theoretical concept of what was due. Tenants were not unwilling to render labour services, which they considered as part of their rent obligation, as long as those services remained within their acknowledged time limits and as long as no other remunerative wage employment was available to them. Customarily the tenant had to work only till noon and was entitled to a light meal.[57] Where these conditions existed compulsory labour either did not become an issue till the settlement operations of the 20th century or fell into oblivion at an earlier stage. The latter case occurred in Manbhum where after 1894 the expansion of coal-mining and the subsequent increase in the demand for unskilled labour had its repercussions on the market for agricultural labour. Under those changed circumstances forced labour in Manbhum became more or less unenforceable.[58] In Ranchi on the other hand compulsory labour constituted an officially recognised grievance at least since 1839, when the then district officer, J. Davidson, commented on it.[59] The whole question had become aggravated when, with the rise of

[55]W. Maude, *Papers on Chota Nagpur Agrarian Disturbances*, Vol. 2, 1890, p. 28.

[56]Ibid., No. 1258-491, LR, 7.4.1890, P. Nolan, SGOB to CCN.

[57]Reid (1912), p. 89.

[58]H. Coupland, *Manbhum Gazetteer*, Calcutta 1911, pp. 147, 150; Gokhale (1928), p. 67.

[59]GOI, LD, March 1869, 14-29: No. 156, 15.3.1859, SAC, L to CCN. Vide attached letter of J. Davidson, dt. 28.9.1839, S13-19; No. 70, 25.3.1859 CCN to SGOB.

Christianity among the tenantry since the 1850s, the raiyats became less subservient to the landlords and in some cases even stopped giving labour services altogether.[60] I shall describe later on how the government tried to cope with this problem.

Bakasht

The Chota Nagpur landlord taxed his tenant in demanding rent, rakumats and compulsory labour, but his most destructive scheme, as far as overall agricultural productivity of the region is concerned, was his success in turning raiyati land into bakasht land.

The term *bakasht* relates to land which is at the absolute disposal of the landlord. It is cultivated either by servants of the landlord, by forced or bonded labour or by share-croppers. I am not certain which was the predominant mode of bakasht cultivation in Chota Nagpur.

In any case, it can be safely stated that the growth of bakasht land resulted in an increasing demand for beth begari. At the beginning of this century, bakasht land amounted to between 10% to 17% of arable land, but it contributed much more than that percentage would indicate to the landlord's income, as has already been mentioned above. Two areas in Chota Nagpur were exceptions to the general picture of availability of bakasht. In South Manbhum and in adjoining Singhbhum bakasht land constituted less than 8% of arable land, whereas in North Palamau and neighbouring Chatra it reached 22% and 28% respectively of arable land. However, from the point of view of agricultural economics it is more important to look at how much of the productive rice-growing low-land was taken up by bakasht. It generally amounted—excluding the exceptions just mentioned—to between 11% and 18% of low-land. The highest concentration of don-bakasht was reached in North Palamau, where more than one-third of low-land (don) had fallen into the hands of the landlords[61] (vide table 2.1, col. 6). B. K. Gokhale wrote on the question of acquisition of bakasht that very few landlords in Manbhum could prove that any portion of their bakasht land was originally reclaimed by themselves or at their own expense.[62] This is true for all other Chota Nagpur districts as well. There can be no doubt, and a number of very knowledgeable settlement officers have agreed on this point, that raiyats in Palamau were forcibly expelled when landlords coveted their land,[63] that it was a common practice of landlords in Hazaribagh to appropriate the best rice lands in

[60]GOB, RD, Oct. 1876, Mis. 4, 82-84: No. 2357, 6.9.1876, W. LeF. Robinson, CCN to SGOB, RD; Reid (1912), p. 90.

[61]Schwerin (1977), p. 144.

[62]Gokhale (1928), p. 41.

[63]Bridge (1921), pp. 57, 118.

each village,[64] that if a raiyat in Palamau improved his holding he would almost certainly be deprived of it by the landlord; and that in West Ranchi the zamindars and their sub-tenure-holders misappropriated all the best lands in the village.[65] One of the landlords' devices to expel raiyats from their rightful land was to increase rent or claim arrears of rent till farming became uneconomical.[66] This of course was strictly illegal, but not much more so than another ploy by which the servants of a landlord cut the crop of a raiyat claiming that the paddy belonged to the landlord as it had been grown on his land. If the raiyat tried actively to defend his property, criminal charges would be brought against him, and if he filed a civil suit to regain possession of his land he was certain to be ruined by the cost of litigation and by the final decree which, in most cases, would go against him.[67] Very often the landlord would not even make use of these pretexts, he simply would make life impossible for his tenants by sending lathi carrying goondas.[68] It may be difficult to imagine the constant and open lawlessness in everything that had to do with the right over land, but it is quite obvious that the British legal and administrative system was unable to cope with the chaotic and free-for-all relationship between tenant and landlord until it finally decided to embark on the costly land settlement operations of the 20th century. This state of affairs, which prevailed up to the first and second decade of this century was hardly conducive to agricultural improvements on the part of the raiyats. For this reason large tracts of potentially arable land, which was normally freely available to the raiyats, remained untilled, although most tenants were in dire need to increase production. They did not, to the extent necessary, level and embank up-land to convert it into Don III. They left most of the irrigation potential of their holdings unused so as not to whet the greed of the landlord!

The Improvident Landlords

The superior landlords on the other hand were similarly in no position to initiate on any meaningful scale land reclamation, irrigation works and so forth. Even though they were potentially extremely wealthy, they were generally, as has been demonstrated by D.C. Wadhwa in his case studies of the zamindars of Tundi and Katras, socially pretentious, improvident and therefore constantly indebted. The lower level landlords, the tenure-holders, were perhaps less improvident, although not less

[64]Sifton (1917), p. 7; *Hazaribagh Settlement Papers*: 18.2.1914, S.D. Sifton SOCN, Note on Agrarian Conditions in the North-West of Hazaribagh District.

[65]SCROB: 11.12.1889, R.H. Renny AC to J.E.K. Hewitt CCN; RD 1258-4911, 7.4.1890, P. Nolan, SGOB, RD to CCN.

[66]Bridge (1921), p. 57.

[67]Hoffmann (1950), p. 2930.

[68]Gokhale (1928), p. 62; Sifton (1917), p. 108.

TABLE 2.2

EXPENDITURES OF PRIVATE ESTATES AND TENURES UNDER PUBLIC MANAGEMENT AS PERCENTAGE OF GROSS INCOME, CHOTA NAGPUR 1873/74-1934-35 (1)

Expenditures	*Court of Ward Estates*	*Encumbered Estates and Tenures*	*Court of Ward and Encumbered Estates and Tenures*
(1)	(2)	(3)	(4)
1. Land Revenue and Cesses	+ 8.1	+ 9.9	+ 8.9
2. Rent and Cesses to Superior Landlord	+ 1.4	+ 5.8	+ 3.5
3. Administrative Expenditures	+ 7.1	+ 6.2	+ 6.7
4. Legal Fees	+ 5.6	+ 4.8	+ 5.2
5. Living Expenses inclu. Education, Medical Expenses, Social and Religious Obligations	+ 8.3	+ 11.3	+ 9.7
6. Improvement of Property	1.7[2] (1.7)	1.0[2] (1.4)	1.4[2] (1.5)
7. Improvements, Charity, Contributions to Schools and Dispensaries, incl. Line 6	3.6[3]	1.7[3]	2.7[3]
8. Survey and Settlement Operations	3.1[4]	1.1[4]	1.8[4]
9. Miscellaneous incl. Line 7 and 8	+ 19.2	+ 9.0	+ 14.4
10. Purchase of Government Securities	+ 13.5	+ 0.6	+ 7.4
11. Repayment and Servicing of Debts	+ 9.0	+ 29.5	+ 18.6
12. Suspense Account and Balance	+ 27.8	+ 22.9	+ 25.6
13. Gross Income (Gross Income in 1000 Rupees)	100.0 (54656)	100.0[5] (48655)	100.0 (103311)
14. Approximate Number of Private Estates and Tenures under Public Management	5-10	36-307	41-317

(1) excluding 1890/91, 1892/93, 1893/94, 1897/98.

(2) pertains only to 1882/83-1889/90, 1891/92, 1894/95-1896/97, 1898/99, 1899/1900, 1915/16-1934/35. Figures in brackets cover the same period as in footnote 3, but include for 1900/01-1914/15 expenditures for charity and contributions to schools and dispensaries.

(3) pertains only to 1882/83-1889/90, 1891/92, 1894/95-1896/97, 1898/99-1934/35.

(4) pertains only to 1873/74-1889/90, 1891/92, 1894/95-1896/97, 1898/99-1914/15.

(5) includes also the income of one to five minor Ward's Estates for the period 1905/06-1929/30.

Report on Wards' and Attached Estates in the Lower Provinces 1874/75-1910/11, Report on Wards', Encumbered, Trust and Attached Estates in the Province of Bihar and Orissa 1911/12-1934/35.

status-conscious, but as their holdings were often small and sub-divided they had to exploit their tenants in order to derive a meagre income from their land.[69] Of course, not all landlords belonged to this poverty-ridden and ignorant category, but the majority of them did. Chota Nagpur could boast of some of the largest estates of the old Bengal Presidency, like Chota Nagpur, Panchet and Ramgarh. The Maharaja of Chota Nagpur, for instance, was lord over 7700 sq. miles. The estate proprietors of the region were, until the Permanent Settlement of 1793, semi-independent petty rulers, and under British rule they had become a well protected and spoiled class. After 1834 these big zamindars and some of their more important tenure-holders were for social and political reasons placed beyond the reach of their creditors. This policy kept the economic position of most old families more or less intact till after 1947. The pampered landlords, especially the estate proprietors and large tenure-holders, had thus no real incentive to improve their property. The result was in most cases an utterly chaotic management, which from time to time was put in some semblance of order when bankruptcy or the minority of the heir brought the estate or tenure under the direct supervision of the district authorities. For these periods, in some cases lasting for decades, annual estate accounts have been published which show clearly (table 2.2) not only the different expenditure pattern for estates under the Court of Wards and Encumbered Estates, but also how much money was reinvested in these estates. The district authorities were forced by government regulations to expend certain parts of the income on "estate improvements". Although they very rarely, if ever, lived up to their own criteria, the authorities were in most cases truer to the image of the "improving landlord" than the proprietors themselves and thus invested comparatively more of the income for productive purposes. Even then, the share of the estate's gross income, which was to some extent spent on irrigation, but mainly on construction of new buildings and their furnishings, remained insignificant during the whole period under review (1882-1934). In nearly half a century the maximum investment for these purposes averaged only 1.5% of the estates gross income. This productive investment compares unfavourably with the expenses for litigation which during the period 1873-1934 consumed 5.2% of the gross income. The administration took another 6.7% share of the income and the living expenses of the proprietors used up 9.7% during the same time period. The unsatisfactory low rate of reinvestment cannot be explained with the land revenue and cess-burden, which was light and averaged only 12.4% of the income (and already included rent to the superior landlord). In the case of the Encumbered Estates heavy indebtedness precluded any diversion of resources towards productive investments. Nearly 30% of the

[69] Bridge (1921), p. 92.

income of the Encumbered Estates was swallowed by debt services and repayments. The estates under the Court of Wards were normally not in such a desperate financial situation, but probably for reasons of convenience investment normally took the shape of purchase of government securities. Whereas less than 2% of the income was spent on improvements, more than 13% went into safe government bonds. From the estate's point of view this was a serious mistake because the money went into conspicuous consumption as soon as the heir came of age.

Besides the majority of privately owned estates Chota Nagpur included also seven Government Estates, covering roughly 10% of the region's territory. These estates were—similar to private estates—of very unequal size, ranging from 4 to nearly 2000 sq. miles in area. Only here the government had permanently stepped into the shoes of a proprietor and zamindar and dealt directly with the raiyats. In 1700 villages of Chota Nagpur, the government could demonstrate how it thought Lord Cornwallis' slogan of the "improving landlord" should be realized. As far as rent assessment and collection is concerned, the government proved to be a more liberal landlord than its private counterpart; the exception to this rule was the Palamau Government Estate where in particular a land settlement of the 1890s played havoc with the rent rates. Similarly more funds were reinvested than in private estates. In the period 1892-1938, 6% of the gross income of the seven government estates were spent on irrigation, buildings, tree planting and other improvements. Although this amounts to more than three times the comparable percentage of expenditure in the estates under the Court of Wards and Encumbered Estates, it is nevertheless disappointing, and belies the official policy statements which stressed the importance of the "raiyats' well-being".

To sum up, an increase of agricultural production which would have made the whole region self-sufficient and would have lifted the majority of holdings at least to a level where they would have provided for the most basic food and non-food requirements, seems to have been impossible throughout the 19th century. The land system of Chota Nagpur which had been fundamentally changed through Hindu immigration from a joint ownership to an hierarchically structured individual ownership system hampered the unfolding of the economic creativity of the people. The peasantry was prevented from improving and cultivating their holdings through irrigation, land conversion and reclamation by rapacious petty landlords, who could only get along by draining the raiyats of their last paisa and by grabbing their best land. If one assumes that the area of low land in Chota Nagpur was a function of the labour invested, one can show that low per capita production of rice as a result of an insufficient per capita endowment with don land was closely related to the greed of the landlords for bakasht land, which in turn is caused by the generally meagre size of the landlords' tenures. When comparing 40 thana-level data relating to the percentage of don-bakasht land and per capita don-endowment it becomes evident that an increasing share of don-bakasht means a decreasing per capita don-endowment.[70] In other words, the peasantry was discouraged from investing its labour in extending and improving its holdings by the successful drive of the landlords for bakasht land. The zamindari class as a whole, be it small or large landlords or even government, could, for lack of means or motivation, not be induced to invest in its own property except on an insignificant scale. These combined obstacles resulted in a level of agricultural production which did not suffice to meet the demand both in regional terms as well as in terms of the needs of the individual peasant.

As is to be expected, such a land system of highly unequal opportunities distributed the scarce agricultural product in a similarly unequal way. The agricultural product, which in our present calculation does not include such valuable assets as lac, tussar-silk, forest products, straw and live-stock, was concentrated, through rent and bakasht cultivation from between 12% (South Manbhum, 1912) and 47% (North Palamau, 1920) in the hands of a landlord minority which, according to the Census of 1931, represented far less than one per cent of the population actively engaged in Chota Nagpur's agriculture.[71] It is quite obvious that such a state of affairs was conducive to social unrest.

Agrarian Tension, Unrest and Revolt in Chota Nagpur

If one considers exploitation, oppression and semi-starvation due to an insufficient level of agrarian production as factors contributing directly to reactions of the suffering victims, then one would expect that the peasantry of North Chota Nagpur, of Palamau and Hazaribagh, should have been the vanguard of an anti landlord movement.

All evidence indicates that the North of the region was economically at a disadvantage when compared to the South and East (Manbhum, Ranchi, Singhbhum). Rice production per head was less and the landlord's share in the overall agricultural product was higher. However, the raiyats of the North reacted only with meek submission to their miserable lot. The settlement officer of Hazaribagh wrote about the peasantry of that district: "It is an entirely wrong assumption to suppose that raiyats will not depose against their own interests. Experience shows that they are commonly forced to do so."[72] It is a fact that in ninenteenth century Chota Nagpur only the peasantry of Ranchi took a defiant stand against its landlords; the

[70]Schwerin (1977), p. 367.
[71]Ibid., pp. 103, 535-538.
[72]Sifton (1917), p. 105.

raiyats of the North rarely even lifted their voice. There is no unequivocal explanation for this phenomenon. The answer lies probably with the ethnic composition of the population. In the Northern parts of the region 50% and more of the people were of Hindu origin. Whereas the extremely stratified rural Hindu society was unable to achieve an economic solidarity against their powerful landlords, which would cut across caste lines, the much more egalitarian tribal communities of the South, although they were in the midst of an important social transformation still retained some of their traditional solidarity, which in former times had produced the powerful armies of battleaxe swinging warriors so much admired by Captain Roughsedge in 1821. As stated earlier, the population share of Hindus is indicative of the length of time they had been infiltrating this once completely tribal heartland. The more Hindus, the earlier the commencement of the immigration, the more time to disrupt the socio-economic fabric of adivasi civilization. In other words, agrarian unrest in Ranchi and its absence in the Northern parts of Chota Nagpur can be explained by the missing potential for solidarity in the Hindu dominated village communities of the North as compared with the tribal society of the South. Only during the course of the 19th century did Ranchi also experience the defeat of the social and economic tribal order, which was accompanied by a lessening of social cohesion which reduced the potential for joint action. It experienced a revolution of the traditional order which had taken place much earlier in Palamau, Hazaribagh and North Manbhum.

The Tribal Land System and the Hindu Tenure-holders

Already the late 18th century saw several pacifying missions of colonial troops in East Ranchi, especially in Tamar, a region almost exclusively inhabited by Mundas. The origins of these troubles are not known, but it is surely no coincidence that the great insurrection which erupted in December 1832 started also in Pargana Sonepur, a part of Tamar. Ten years earlier, a new Maharaja of Chota Nagpur had inherited the throne at the age of 19 and had started giving temporary grants of villages to "foreigners", that is Dikus, in lieu of cash payments for the purchase of their merchandise. The Diku tenure-holders dispossessed several of the Mankis (circle headmen) which sparked off a large-scale armed uprising against the Dikus and, in the end, against the British troops who came to the rescue of "law and order". One of the tribal leaders put the popular complaint in a nutshell in saying: "They (the Dikus) have taken away from us our trees, fishes, lands and jagirs. They lend us one rupee, but take three from us."[73] This was to remain valid for the rest of the

[73]J.C. Jha, *The Kol Insurrection of Chota Nagpur*, Calcutta 1964, p. 151.

century. Government reaction culminated in a new administrative set-up, which supposedly brought British administrators into more intimate contact with the people.

In 1845, four German Protestant missionaries settled in Ranchi, and five years later they baptised their first four Oraon converts. This little-noticed episode proved highly significant during the second half of the century. At the time of the Mutiny in 1857 the Christian community was still rather small, but large enough to have already attracted the scorn of some of the landlords, who in 1855 had conducted a concerted action against converts in thirty villages.[74] During the Mutiny, when the four infantry detachments of the Ramgarh Batallion revolted and the authorities fled the administrative headquarters, some of the landlords took the opportunity to again persecute the Christians who were compelled to seek the protection of the jungle.[75] It is improbable that religious intolerance on the part of the landlords prompted these persecutions; they were more afraid of the reasons for which the young converts had embraced the new faith. All missionaries agreed that spiritual benefit was of little import to the new members of the church, but that the conversion was seen by the Oraons and Mundas as an act of emancipation from Hindu oppression.[76] For the Christians of the first generation the church was nothing more than a new vehicle in an old struggle. It offered a new framework for a village transcending solidarity which had been damaged by the successful Hindu penetration.[77] It fits well into this picture that in the two first periods of mass conversion of 1858 to 1860 and 1868 to 1870 whole families, clans and villages converted together, led by the most important families of the community "who still had something to lose and who at the same time possessed some kind of economic resistance". The landless or small raiyat remained uninterested in adopting Christianity.[78] Again, it is surely no coincidence that after the first success in the Oraon settled parts of Ranchi, the main movement towards the new religion developed in the Munda heartland of the district,[79] which had been the scene of unrest and dissatisfaction since the late 18th century. In the third period of mass conversion between 1885 and 1889, this time to the Catholic

[74]L. Nottrott, *Die Gossnersche Mission unter den Kolhs*, Halle 1874, p. 198.

[75]GOB, JD, Apr. 1859, 77-79: No. 156, 15.3.1859, J.S. Davies, SAC, L to E.T. Dalton, CCN.

[76]Th. Jellinghaus, "Die Kolhs in Ostindien und ihre Christianisierung" in: Allgemeine Missionszeitsohrift, Guetersloh 1874, Vol. I, p. 24, p. 182; T.S. Macpherson and M. Gttallett, *Ranchi District Gazetteer*, Patna 1917, p. 224.

[77]GOB, RD, July 1872, 33-35: No. 1354, 30.5.1872, E.T. Dalton CCN to H.L. Dampier, SGOB, R.D.

[78]Jellinghaus (1874), pp. 182, 184, 264; L. Nottrott, *Die Gossnersche Mission unter den Kolhs, 1874-1887*, Halle 1888, p. 41.

[79]Jellinghaus (1874), p. 213; Nottrott (1874), p. 189.

Church under the energetic leadership of Father Lievens, S.J., the tribal population itself called the movement towards Catholicism "Rai", that is alliance against the zamindars.[80] I therefore believe it to be quite clear that the large scale adoption of Christianity was motivated by the population's anxious desire to obtain a new ally against the landlords in the supposedly powerful white man's church. The conversion was caused by agrarian unrest, but it was not at the bottom of the conflict as the landlords liked to maintain thus turning the issue upside down.

After the Mutiny there was a backlash against some landlords. Their bakasht land was forcibly seized by their Christian tenants, who claimed that the land had originally belonged to them.[81] The struggle revolved especially around a certain class of land called "Bhuinhari" which by tradition was only burdened, if at all, with a minimal rent obligation and constituted a remnant of the old khuntkattidar rights of the descendants of the village founder. Some British officials had already in previous years sensed the prime importance of the Bhuinhari question and had urged a general registration of those lands (Davidson in 1839, Ricketts in 1855).[82] The Mutiny had alarmed the government, but the necessity for a measure of this kind was seen only after the clashes of 1858 had occurred. The ensuing registration proved to be partial, covering only a limited area. Although this first Bhuinhari registration succeeded in those villages covered by it to a certain extent in reducing tension over land questions,[83] frictions in other parts of the Ranchi district continued and slowly built up to a climax. The constant deprivation, exploitation and oppression of the tribal tenants found its temporary outlet in what was to be called the "Sardar Larai", the War of the Leaders.

The Sardar Larai

This social movement commenced in the 1860s and stayed for most of the time of its existence within the narrow bounds of law and order. Its main weapon was the petitioning of government and later on rent refusals and a landgrab movement which, of course, was illegal. Its leaders, the sardars, belonged mostly to the traditional tribal elite and at one point in time had all been members of the Protestant mission.[84] As pointed out earlier, whole families, clans, and villages had taken refuge with the missionaries at the same time. When asked to elect church elders they had chosen members of the established leading families, who, because of their

[80]Hoffmann (1950), p. 3843.

[81]GOB, JD, Apr. 1859, 77-79: No. 156, 15.3. 1859, J.S. Davies SAC, L. to E.T. Dalton CCN.

[82]GOB, JD, Apr. 1859, 77-79: No. 70, 25.3.1859, CCN to A.R Young, SGOB.

[83]GOI, LD, Mar. 1869, 14-29: No. 227, 31.1.1868, CCN to SGOB.

[84]Jellinghaus (1874), p. 256; Nottrott (1888), p. 35.

discontent with the existing order of things, had been the driving force towards the new religion.[85] The first petition originating from Christian tenants which we know of was initiated by those elders in the spring of 1864. They tried unsuccessfully to hand over the document to the visiting Anglican Bishop Cotton, who was mistaken for the Governor-General of India. The leading missionary of that time prevented its presentation by tearing it up.[86] The elders, not easily to be discouraged, started the first of many collections of money and in 1865 offered half of the collected money of Rs. 300 for the construction of a seminary building, if the missionary would be willing to take up a lawsuit on their behalf in Calcutta with the rest of the money. The final rupture with the missionaries for this first generation of elders came in 1866 when the leading missionary hit two of them and excommunicated 25 others. These two dozen ex-elders, whom one missionary called the "Fathers of the Christian community", have to be regarded as the first sardars.[87] For one and a half decades they were, under the leadership of Manwassi Munda, who died in 1887, the heart and backbone of the growing movement. In 1879, the Sardar Larai had become such a headache to the Protestant mission that it found itself compelled to disassociate the church from a good number of its young intellectuals, its catechists, teachers and candidates for the ministry.[88] The ex-catechist Johann became Manwassi Munda's successor. Both men belonged to the Munda parish Govindpur to the South-West of Ranchi. It is interesting to note, but not really surprising, that 21 out of the 36 signatories of the numerous petitions of the year 1884 were inhabitants of pargana Basia and Sonepur, that is to say of the two Munda parishes Govindpur and Burju where the majority (69% in 1888) of protestant converts lived.[89] The less numerous Oraon converts did participate in the movement, but, as one of the local officials put it, "being a more practical people, have grown tired of paying money for nothing and dropped the business altogether."[90] The German missionary, Nottrott, dates this break between the Oraons and the movement from 1887, when he sued two Oraon Sardars for libel. Nottrott claims that this prosecution destroyed more or less completely the Sardars' reputation in the eyes of the Oraon community.[91]

The Sardar Larai started in earnest with the excommunication of the "twentyfive" in 1866. The ex-elders reacted by collecting the first really

[85]Nottrott (1888), p. 106.

[86]Jellinghaus (1874), p. 253.

[87]Ibid., p. 256.

[88]Nottrott (1888), p. 30.

[89]A. Nottrott, "Der gegenwaertige Stand der Kols-Mission mit besonderer Beruecksichtigung der Landfrage" in: *Allgemeine Missionszeitschrift*, Guetersloh 1889, vol. 16, p. 257 ff. p. 306.

[90]SCROB: 10.3.1898, A. Forbes CCN to Bolton with attachment, 8.1.1898, Note by H.C. Streatfeild DC, L.

[91]Nottrott (1889), p. 270; SCROB: 27.1.1892, A. Nottrott to DC, L.

substantial amount of money (Rs 7500) to take up the legal fight for their lost traditional land rights on behalf of the whole tribal community, Christians as well as non-Christians. In December 1866, a delegation of approximately one hundred peasants went to Calcutta to see the lawyer, G.M. Tagore, who advised them to petition the Commissioner of Chota Nagpur and, if that turned out to be of no avail, the Lieutenant Governor of Bengal.[92] The two petitions of January and September 1867 list the following main complaints: appropriation of raiyati land by landlords, their refusal to grant rent receipts, illegal rent increases and, as demands: registration of land, fixation of rent rates and the possibility to pay rent directly into the district treasury instead of to the constantly cheating and oppressive landlord, who, however, by law was the sole person entitled to receive rent.[93] "Direct rent payments" remained a favoured battle cry of the movement, which goal however was unattainable under the ownership pattern as established by the Permanent Settlement of 1793. What the Sardars really asked for was a reversal of the Permanent Settlement. For obvious reasons this turned out to be a political impossibility under colonial rule. Only the post-indepedence land reform measures fulfilled this point on the Sardars' list of demands. They were more successful with their demand for registration of land, which was carried through as a "political necessity" with two important limitations from 1869 to 1880. The land registration and restitution of illegally seized land was confined to Bhuinhari and Manjihas (=Bakasht) and thus left undemarcated and undefined most of the arable land which was called Rajhus. Contrary to Bhuinhari, Rajhus land paid full rent rates, which were of course, a major bone of contention between landlords and tenants. The registration, which comprised 2482 villages and was considered a major undertaking by government also in its cost implication (Rs 270,000), therefore did not live up to the expectations, neither of government, which had hoped for peaceful docility afterwards, nor of the tribal tenantry, which saw the major class of land, Rajhus, still open to the landlord's inroads.[94] The result was a continuation of the agitation which had to be faced by British officials who were convinced that they had already done everything possible by launching the Bhuinhari Survey. The next ten years after the survey witnessed what some officials called "petition mongering". Petitions, complaining about the implementation of the Bhuinhari Survey, and demanding a new survey and "direct rent payment", went twice to the Governor-General in Calcutta (1881, 1886) and twice to the Secretary of State for India in London (1882, 1888); a large number of petitions were handed to local officials and missionaries.

[92]GOI, LD, Mar. 1869, 14-29; No. 227, 31.1.1868, CCN to SGOB.

[93]GOI, LD, Mar. 1868, 14-29: 21.9.1867, Petition to LGOB.

[94]*Papers relating to Chota Nagpur Agrarian Disputes*, Vol. 1, 1890: 25.11.1880, Resolution GOB, RD.

The reply was always negative, leading to a partial change of tactics on the part of the Sardars by the mid 1880s, which was supposedly more harmful than beneficial to the trusting raiyats and the tribal order it tried to serve. In June 1886, the Commissioner of Chota Nagpur was petitioned to grant the right of direct rent payments to the district treasury till the government had decided on previous petitions. This was followed up by a no-rent-to-the-landlord campaign which reached its first climax in 1887[95] and its second one in 1889.[96] The peasants either withheld the rent completely or took advantage of a law of 1862 which allowed temporary rent deposits in the district treasury. Those last cases jumped from 1460 in 1886 to over 2000 in 1887 and 1888, and to nearly 4000 in 1889.[97] In both years 1887 and 1889 the Sardars also recommended to the raiyats to repossess land from which they felt they had been forcibly evicted by the landlords. The authorities reacted promptly by strengthening the local police force and travelling magistrates who, especially in 1889, passed sentences so indiscriminately that a good part of them had to be voided in courts of appeal. It seems that the no-rent campaign had devastating effects, because, if the rent was due from a village or circle headman, his "tenure" was put up for auction, thus destroying another remnant of the dying tribal order. In case an individual tenant withheld rent a similar fate awaited his holding. The Sardars countered these defeats with the consoling advice that all this and more would be recovered after the impending total victory over Hindu domination.

After nearly a quarter of a century of agitation the Sardars had reached the end of their tether. They had strained their imagination and tried nearly everything except violence. Some of them had grown prosperous by the constant money collecting drive meant to defray the expenses of their legal counsels in Calcutta.[98] Over the years the district authorities had tried several times to put an end to these collections which went against the public interest in its official interpretation and, as the officials had claimed, served only the personal enrichment of the Sardars and their lawyers.[99] One of the lawyers, E. Jacob, who was to serve as the counsel of Birsa, remarked in 1898 that he had seen money receipts for over Rs. 260,000, which the Sardars had paid their lawyers in Calcutta since 1883.[100] It seems certain that substantial sums had passed from Ranchi into the hands of very often fraudulent and completely irresponsible third

[95]SCROB: No. 1 T-J, 19.11.1887, C.C. Stevens CCN to SGOB.

[96]SCROB: No. 1258-491 LR, 7.4.1890. P. Nolan, SGOB to CCN.

[97]W. Maude, *Report on Agrarian Disturbances in Chota Nagpur*, 1890, Vol. II, Attachments C and D.

[98]Nottrott (1889), p. 260.

[99]SCROB: No. 476 R, 17.7.1884, J.F.K. Hewitt, CCN to SGOB; No. 1 T-J, 19.11.1887 C.C. Stevens CCN to SGOB.

[100]SCROB: No. 407, 13.8.1898, H.C. Streatfeild, DC, L to CCN.

rate advocates. Some of this money had clung to the pockets of the Sardars and it is quite possible that several "leaders" had played the rupee-game with their gullible countrymen.

Everything had been tried except force. The explanation for this abstention is probably fourfold: (1) the belief of the population in the good intentions of the government, (2) the trust in the mediation of the powerful white missionaries, (3) the restraining influence of the missionaries and (4) the very reasonable fear of the modern weaponry of the Indian army (the regular police was unarmed) which had to be matched on the tribal side by the traditional bow and fighting axe. Even so, there were always rumours about the possibility of a general insurrection. The missionaries had warned of an uprising in their great petition to the Lieutenant Governor in 1876.[101] Eleven years later people said that Queen Victoria (!) had been asked for permission to fight the Sahibs. The Protestant missionary Nottrott even knew the appointed day in 1891 and claimed that only internal differences had prevented the outbreak of the long expected event,[102] which finally occurred at Christmas 1899, under the leadership of Birsa.

The Revolt of Birsa

Birsa's rise to the pinacle of the agrarian movement had been comparatively swift. He started to attract attention in 1895 and died in jail in 1900. In between these dates he was imprisoned for more than two years for fostering public unrest. His social background was humble. Contrary to earlier tribal leaders, he was not the scion of an elite family. His father was a landless labourer or share-cropper and sometime catechist of the protestant and then the catholic mission. Comparable to the younger generation of Sardars Birsa had a rudimentary knowledge of the three "R's", which he had acquired during three years at a mission school. He was born when the Sardar Larai was already in full bloom, reached adulthood when it had proven futile, and brought it to its bloody yet inspiring end when he himself was only in his mid twenties.[103] Birsa propagated a special blend of political demands and religious teaching. He was indebted to the Christian doctrine for many of his views. However, he was not an entirely new phenomenon in Chota Nagpur. In fact he was preceded by at least two similar prophets, Mahsihdas Birsa in the 1870s and "Johan the Baptist" in 1881. Just like Birsa, Johan the Baptist used Doisa, a former palace of the Maharaja of Chota Nagpur, as a platform for his call for the

[101] *Papers relating to Chota Nagpur Agrarian Disputes*, Vol. 1, 1890: 17.5.1876, Petition D. Onasch to LGOB.

[102] SCROB: 27.1.1892, A. Nottrott, G.E.L. Mission to DC, L.

[103] S.K. Singh, *Dust Storm and Hanging Mist* (A Study of Birsa Munda and his Movement in Chota Nagpur), Calcutta 1966, pp. 31-38.

"RAJ". Raj in this connotation stands for Adivasi rule and supremacy. It is, by the way, interesting to note the importance tribal leaders attached to the few buildings like the Nawrattan Palace at Doisa (built 1687-1711) as symbols of their golden age, although they probably only indicated Hindu ascendency in that part of the country.[104] Birsa dreaming of reviving this glorious past, as did many tribal leaders before him, understood the protection the colonial rulers afforded to the oppressive land system he was to fight, and resolved on a general insurrection. That he was able to succeed in this endeavour, which had been rumoured about since at least 1887, is partly due to his commanding personality and talents, but even more so to the general exasperation of his people after fourty years of more or less non-violent and very costly agitation. What I wish to stress here is that Birsa was a product of the general situation of his time; a man to win hearts, but picking up the bits and pieces and ideas where his predecessors had left them. Birsa was not a towering innovator, he was the coping stone of a building, which immediately afterwards crashed only to form the basis for social and political movements which were to start after independence.

The problem of Chota Nagpur's tribal population during the last century have to be viewed on two levels: one is the loss of its material basis, the other is a crisis of identity. Whereas the Sardars attacked the first problem as the most pressing, they had not been able to give even a theoretical answer to the second set of problems. Here is where the three missions stepped in and, of course, Birsa with his two predecessors. Men and leaders like Birsa were not exceptional in India. S. Fuchs has listed dozens of examples of this kind of leadership among Indians. The loss of identity, difficult to quantify, but nevertheless real, results from a clash of cultures, which "took place", as S. Fuchs wrote, "nearly always in colonized countries which were economically underdeveloped, highly isolated, politically acephalous. In India, prior to European powers, we do not speak of 'colonizers' and the 'colonized', but the effect of the high-caste Hindu and Muslim rule on the low castes and aborigines was practically the same as in other countries colonized by Europeans."[105]

The confrontation between the Hindu and Adivasi civilization, heightened during the 19th century by colonial courts, legislation and police, which were effectively manipulated only by Hindus, led not only to the disintegration of the traditional tribal order and to material disadvantages for the tribes, but also to a break-up of the feeling of communal togetherness in their villages. Strife and tension separated not only different ethnic groups, but also clans and families belonging to the same tribal community.

[104]Nottrott (1888), pp. 166, 176; SCROB: No. 1 T-J. 19.11.1887, C.C. Stevens CCN to SGOB.

[105]S. Fuchs, *Rebellious Prophets*, London 1965, p. 2.

Villages broke up in factions, the stronger trying to adopt themselves to the new circumstances by disregarding the rightful claims of the weak. The possibility of escaping to the tea plantations of Assam did the rest; the destruction penetrated into the core of the family. The tribal population, formerly renowned for its truthfulness and dignity, now tried the easy way out and bent the truth. It hoped to gain the future by disclaiming Sing-Bonga and by opting for the Hindu pantheon and Jesus. Only a few, however, succeeded in making peace with the present. The majority still finds itself at the fringe of society. The tribal population has been victimized by a situation which cannot be mastered on an individual basis alone. The lack of self-confidence is suffocating, only the people's trust in their own ability to succeed and to organize will bring about a solution. Jean Paul Sartre has rightly pointed out that the status of being a colonized native is a neurosis which has been introduced and is kept up by the colonizers *with the consent* of the colonized. It is necessary to break this consent. Birsa, and that is his importance, did just that, even if only for a very brief moment in time. With some justification therefore Birsa has risen in post-independent India to the stature of a regional hero and has become a symbol for the future of a viable adivasi culture. Today, his bronze statue stands in front of the main gates of the Heavy Engineering Corporation at Ranchi: The revivalist Birsa as the patron of the future; a contradiction which is, however, only a superficial one.

The Government's Response

The colonial rulers, like most governments, were slow to react to the problems they were forced to face in Chota Nagpur. The British, by their comprehension of what was politically feasible for them in India felt compelled to remain within the framework of the Permanent Settlement of 1793. Thus their hands were tied, and the process of agrarian reconstruction in Chota Nagpur which had to be initiated by abolishing the landlord system as the first most needed step, had to wait till after independence. Under these circumstances, the British were not able to offer more than merely piecemeal solutions and even these were arrived at very slowly and implemented at an even slower pace. The Provincial and Central Governments in Calcutta (since 1911/12 in Delhi and Patna) moved on three fronts with all the snail-like cumbersomeness of a huge bureaucracy which is under no direct pressure from a voting public. These three fronts were first an expansion of the administrative system in Chota Nagpur, second, survey and settlement operations, and third, legislative measures.

Development of the Administrative Structure

The first British reaction was to create over several decades an increasingly tighter administrative network. The process started with the establishment of the far-flung Ramgarh District in 1780, part of which (the later Manbhum) became temporarily included in the Jungle Mahals in 1805. The so-called Kol-Rebellion of 1831/32 led directly to the constitution of the South West Frontier Agency which in 1854 was reconstituted as the Division of Chota Nagpur. The Agency included most of the former Ramgarh District plus Singhbhum and the Manbhum part of the Jungle Mahals. The Agency was, however, when compared to the former Ramgarh District, administered by a greatly increased staff, led by the agent and his four principal assistants, who were later to be called Commissioner and Deputy Commissioners. In 1852 the first two subdivisions were already created for Palamau and Gobindpur (Dhanbad) as part of the Ranchi and Manbhum District respectively. This was followed in 1870 by the establishment of the Pachamba (Giridih) Subdivision in Hazaribagh. This became necessary because of increased coal-mining due to the opening of the first railway link in the division. In 1891 the Palamau Subdivision was elevated to the rank of a district, but otherwise no significant changes in the administrative set-up occurred for thirty years. Only the Birsa Revolt jolted the administration into new activism. Two years after Birsa's death in jail the Gumla Subdivision was founded in 1902, followed by Khunti in 1905,[106] thus bringing public administration, courts and hopefully justice much closer to the most troubled parts of Ranchi.

Survey and Settlement Operations

The second possible British reaction to the agrarian problems of Chota Nagpur were the so-called Survey and Settlement Operations during which the land was measured and mapped and the various rights in land recorded. This was, however, a costly and legally complex proposition which until 1902 was resorted to only in the cases of estates owned by the government or administered by the Court of Wards or even more rarely in estates which had come under the Encumbered Estates Act. In 1902 the first district-wide settlement operations were started in Ranchi, as a direct consequence of the Birsa Revolt. The settlement was an extremely important event and men like the Jesuit and committed social scientist Hoffmann, who knew Chota Nagpur intimately, became rather poetic in describing its results. Hoffmann wrote of the Ranchi settlement of 1902-1910 that "it literally resembled a calm, sunny morning after a long hurricane, which had reduced

[106]Later on several more subdivisions were constituted: Chatra 1914, Simdega 1915, Dhalbhum 1920, Latehar 1924.

all the Mundas to extreme poverty and had thrown hundreds of thousands forever far away from their own dear country. For those remaining, it brought peace and security".[107] A general settlement of Ranchi, comprising all classes of land and not just Bhuinhari and Manjihas (Bakasht) land as in the 1869-80 survey, had been a continuous and unanimous demand by officials and non-officials alike for at least a quarter of a century.[108] The superior government, in principle not opposed to it, had however always countered this proposal by pointing to its high costs.[109] The strongest argument to overcome this cost-impediment had in the end been the tribal insurrection of Christmas 1899. As in most cases of non-democratic governments, the British proved better able to respond to the voice of violence than to rational demands.

Before continuing the discussion of the results of these district-wide survey and settlement operations, which took nearly a quarter of a century (1902-1925) to complete, we should turn first to the third option available to the British in combating agrarian unrest and social tension, i.e. to legislation.

Agrarian Legislation

In Chota Nagpur the Government was confronted with one basic question: Should it treat this region as a special case in keeping with its social and economic development or accord it equal rights and duties similar to those of other parts of the Bengal Presidency? Very early the British had treated Chota Nagpur differently from the rest of Bengal. Whereas, for example, estates under the Permanent Settlement in Bengal proper were assessed to 90% of the income received from rent, in Chota Nagpur the land revenue bore, with the exception of Panchet in Manbhum, no relation to the rent roll of the estate, but was more a kind of a mutually agreed tribute, which the local raja presented to the conquering British. This exceptional treatment of the province was once more sanctioned with the constitution of the South West Frontier Agency as a non-regulation province in 1833. Creditors, for example, were no longer able to obtain sale decrees for the property of defaulting estate proprietors and tenure-holders. Landed property had been made unsalable for fear of otherwise threatening agitation and insurrection by the people in favour of the old

[107] Hoffmann (1950), p. 2404.

[108] *Papers relating to Chota Nagpur Agrarian Disputes*, Vol. 1, 1890: 25.11.1880, Resolution GOB, RD; SCROB: NO. 15 T-J, 30.11.1889 W.H. Grimley, CCN to SGOB; 10.3.1898, A. Forbes CCN to Bolton; No. 1 T-J, 19.11.1887, C.C. Stevens CCN to SGOB; 24.9.1890 F. Hahn G.E.L. Missionary to S. Bailey LGOB; 6.11.1892 Pertap Udia Nath Sahi Deo Maharaja of Chota Nagpur to C. Elliot, LGOB.

[109] W. Maude, *Report on Agrarian Disturbances in Chota Nagpur*, Vol. 2, 1890: No. 1258-491 LR, 7.4.1890, P. Nolan SGOB to CCN.

proprietor.[110] In later years it was reasoned that the traditional landlord proved less rapacious, if only because of his inefficiency, than the probably moneylending successor. The district administration thus became burdened with the management of estates and tenures whose proprietors could no longer pay their debts. The property remained in public custody till all debts had been cleared off, whereupon it was handed back to the proprietor, who in most cases just started the debt-circle all over again. When shortly before the reform legislation of 1876[111] this system was most abused, there were in the division an average of 307 estates and tenures under public administration[112] with an average yearly rent roll of Rs. 393 per estate or tenure.

In many important aspects Chota Nagpur remained a non-regulation province till Independence, requiring a separate legislative procedure. This was especially true when legislation touched landlord-tenant relations. The first law in Bengal that tried to codify those relations was the Bengal Rent Act (Act X of 1859). It was never quite clear whether the act was legally in force in Chota Nagpur or not. The Board of Revenue recommended to follow the spirit of the law.[113] In the end it was decided to play safe and to re-enact the law with some modifications as the Chota Nagpur Landlord and Tenant Procedure Act (Bengal Act I of 1879), which remained unaltered and in force until 1903. The Rent Act tried to deal with the main *casus belli* between the two contending parties, that is the question of rent enhancement. The most important stipulation stated that "private enhancements" solely between landlord and tenant were illegal. A rent increase was binding only if the district officer had been a party to it by issuing a written notice of enhancement on the application of the landlord specifying the grounds on which an enhancement had been demanded and granted.[114] If the tenant objected a suit for rent enhancement would follow. This was the only legal way to enhance rent after 1859. It is perhaps not surprising that only the most minimal use was made of these rather cumbersome provisions. The increasing number of rent suits in Ranchi and Manbhum, which I have mentioned earlier, were overwhelmingly suits for arrears of rent and not for an increase of rent rates. By claiming arrears of rent calculated at enhanced rates which had never before been collected, or

[110]The case in point was the attempted sale of the Panchet Estate, Manbhum, in 1795.

[111]*Chota Nagpur Encumbered Estates Act* (*Act VIII of 1876*).

[112]An additional 10 estates and tenures were administered by the Court of Wards with an average annual rent roll of Rs 38,860 per estate and tenure.

[113]SCROB: No. 3115, 6.11.1869, E.T. Dalton CCN to A. Eden SGOB, J.D.

[114]*The Bengal Rent Act* recognized three justifications for enhancing rent. A general increase of prices was not one of them and appears first in the *Bengal Tenancy Act of 1885*. One of the modifications of Act I of 1879 was to empower the district officer to decide the issue of a rent enhancement "as to him may seem fair and reasonable" without binding him to any predetermined criteria.

by sueing for arrears of rent on land, for which rent had never before been paid, the landlords were able to obtain court decrees which not only forced the tenant to pay arrears which he had not incurred, but also unwittingly legalised new, increased rates of rent. It is no great consolation that in the second half of the 19th century even in the more "civilized" province of what is nowadays Uttar Pradesh, the situation as far as rent suits were concerned was rather similar to that of Chota Nagpur.[115]

The rent act of 1879 incorporated certain provisions based on the custom of Chota Nagpur. Rent enhancement for land owned by Bhuinhars of Khuntkattidars became limited and rent rates of newly reclaimed rice-land (korkar) were regulated by local custom. It was further stipulated that the court on application of either the tenant or the landlord could commute the tenant's obligation to forced labour (beth begari) into a money-addition to the rent. To demand forced labour in excess of the traditional norm was a punishable offense. As far as the issue of forced labour was concerned, the rent act of 1879 signified a great step in the right direction when compared with the Chota Nagpur Tenures Act (Bengal Act II of 1869) of a decade earlier. At that time the government's main concern was the restitution of illegally seized Bhuinhari land to the former tenant owners and the registration of Bhuinhari and Manjihas (Bakasht) lands. Under the act of 1869 only Bhuinhars and the landlords could demand a commutation of the forced labour obligation. Other classes of tenants, contrary to the rent act of 1879, were not permitted to apply and the extraction of excessive forced labour did not become an offense, although this had been discussed in the Legislative Council at that time.

The commutation clauses of 1869 and 1879 showed no appreciable impact. Only 1167 applications for commutation were filed after 1869 and they resulted mainly from the exhortations of the Indian officials responsible for the Bhuinhari Survey. During the nine years following 1879, the courts received not more than 95 (!) requests for commutation. These poor results were probably due to (1) the innate fear of the tribal tenantry of the legal machinery which would have to be employed, (2) the non-involvement of the Christian tenants who in many cases had already unilaterally ceased to render labour services and (3) the fact that in many regions it remained more convenient to offer labour services rather than payment in cash. Contrary, however, to what one would deduce from the very limited response to the opportunities offered by the commutation clauses, the question of forced labour remained a burning issue in Ranchi and was, according to the conviction of the government one of the main reasons for

[115] T.R. Metcalf, *The Aftermath of Revolt, India 1837-1870*, Princeton 1964, p. 183.

the disturbances of 1889[116] mentioned earlier. As a belated response Calcutta finally passed the Chota Nagpur Commutation Act (Bengal Act IV of 1897), which allowed commutation without prior application of one of the two parties, if it was deemed necessary "for the preservation of peace in the country".[117]

Thus after many trials and errors it was acknowledged that under the prevalent social and economic conditions of Chota Nagpur it was futile to expect the tenantry to take the initiative in court proceedings against the landlords. The government applied this new instrument in the land settlement operations, which began in 1902 in Ranchi. With its commutation into a money addition to rent forced labour slowly ceased to be a problem in Chota Nagpur, although landlords, especially in Palamau and Hazaribagh, often succeeded in exacting labour services after they had been commuted. Some twenty-five years later it was said, however, that in Ranchi forced labour had, with some exceptions, ceased to exist.[118]

Another problem which was not an issue with the population itself but with the officials, was the increasing sales of raiyati land (as evidenced by the registration statistics which showed a more than sixfold increase in fifteen years).[119] It had been estimated in 1899 that approximately three quarters of the Bhuinhari land that had been registered between 1869 and 1880, had already passed from the hands of the original owners into those of outsiders, mainly moneylenders.[120] It was feared that a similar fate was in store for the lands of the autochthonous population after it had become more valuable through registration and mapping during the settlement operations. The government was afraid of the social and political consequences in case the tribal tenantry should be depressed to a share-cropper status under Hindu tenants or reduced to the status of landless labourers. After decades cf discussion Calcutta had finally in the 1890s arrived at the conclusion that it was necessary to prevent transfers of raiyati lands by prohibiting them altogether. The Central Provinces Tenancy Act of 1898

[116]W. Maude, *Report on Agrarian Disturbances in Chota Nagpur*, Vol. 2, 1890: No. 1258-491, LR, 7.4.1890, P. Nolan SGOB to CCN.

[117]GOI, LD, Oct. 1897, 6-24: 2.1.1827, W.H. Grimley, Statement of Objects and Reasons.

[118]Sifton (1917), p. 122; Bridge (1921), p. 92; F.E.A. Taylor, *Ranchi SR*, Patna 1940, p. 52.

[119]Registered sales of raiyati land in Chota Nagpur:

1883/84—1887/88	3525
1888/89—1892/93	10065
1893/94—1897/98	17983
1898/99—1902/03	22235

These figures show undoubtedly a rising trend in sales, even if registration became more popular over time. Twenty Years Statistics, 1883/84-1905/06, 23 Vols. Calcutta 1904-1906. Vol. Registration.

[120]SCROB: No. 713 J, 14.1.1899, A. Forbes CCN to SBR, GOB.

embodied this decision for the first time followed by the famous Punjab Alienation of Land Act of 1900.[121] The Chota Nagpur Tenancy (Amendment) Act (Bengal Act V of 1903}, which revised the rent act of 1879, did the same for Chota Nagpur. Only a court could now order the sale of land for arears of rent or for repayment of government loans. Otherwise land could only be leased for a specified limited period of time. These protective measures were well intended but did not work as expected.[122] They rather lowered land prices to the disadvantage of the raiyat.[123] Consequently these clauses were successively relaxed in 1920 and 1938. After 1938 only members of tribal groups had to seek permission of the district authorities if they wanted to transfer land.

The major work of legislation was undoubtedly the Chota Nagpur Tenancy Act (Bengal Act VI of 1908) which consolidated and superseded all previous agrarian legislation for the province. While in 1903 the Khuntkattidars, the original joint ownership community, had been recognised and protected by the law, now the status and the land of the traditional village headman came under the protection of the courts. The introduction of new levies (abwabs) and the reintroduction of forced labour (beth begari) as part of the rent demand were disallowed, and the commutation of rent in kind into a cash rent was emphasized once more.

If one reviews British agrarian policy and legislation in Chota Nagpur up to 1947, a shift of focus is immediately discernible. While in the first one hundred years of British rule the landlord class received most of the attention and favourable treatment making, for example, their estates and tenures virtually unsalable, the tenantry gained in political importance during the rest of the time. It is not difficult to find an explanation for this shift. Due to the nature of the Permanent Settlement the estate proprietors slowly lost their importance as the prime source of state revenue. In addition, with the consolidation of British power, they also lost their potential as trouble-makers. In a way, they were effectively bought off by Lord Cornwallis' ingenious measure of 1793. On the other hand, apart from the risings in 1831/32 and 1899, which were motivated by social and economic fears, which did activate the government, the whole perception of society by British officials changed from an 18th century aristocratic outlook to one effectively altered by the industrial revolution.

Although Chota Nagpur as a whole was always accorded a different treatment from the rest of the Bengal Presidency, it was only from 1869 onwards that the British Indian government legislated especially for this

[121] D. Rothermund, *Government, Landlord, and Peasant in India*, Wiesbaden 1978, p. 74ff.; 177f.

[122] Taylor (1940), pp. 56, 91; Appendix XXXIX; Lacey (1942), Appendix p. X; Gokhale (1928),p. 128

[123] *Bihar and Orissa, Legislative Council Proceedings*, Patna 1920, Vol. 6.

region and tried to recognise and protect existing customs by legislation and through the courts. In 1869 the special status of Bhuinhars was recognised by registering their lands; ten years later it was decided that the maximum rent rates of land, owned by Bhuinhars or Khuntkattidars, could only be 50% of equivalant raiyati land. Rents for newly reclaimed riceland (korkar) had to be based on prevalent custom, and extortion of forced labour in excess of the traditional norms of approximately 14 days was expressly prohibited. Finally in 1908 the status of village headmen and khuntkattidars became recognised in all its complexity for the first time. Custom, if not running contrary to the law, was declared to be admissible in a court of law. While these developments were favourable to the tenants, they came much too late. The traditional institutions and privileges which were now sought to be protected, had survived, if at all, only in isolated pockets. In most part of the province they had already been destroyed by the inroads of the Hindus with the assistance in later years of the legal machinery of the British courts. This part of the agrarian legislation though important as a belated acknowledgement of tribal rights, was nevertheless nothing more than a rearguard action. However, measures like the commutation of forced labour (1897) and like the commutation of rent in kind into cash rents (1908) or the prohibition of the transfer of raiyati lands (1903) have to be viewed differently. They were socio-economic legislation which tried to change and not to preserve. Into the same category belongs the Bihar and Orissa Kamiauti Agreements Act (Act VIII of 1920) which sought to improve the lot of bonded labourers or Kamias—an outrage, which is still (1976) widespread especially in Palamau and Hazaribagh.[124] Generally speaking this kind of legislation remains a mere declaration of intent and is an unenforceable ornament of the statute book, if the prevalent socio-economic climate does not support it. For this reason the final commutation of forced labour succeeded fairly well, whereas the prohibition on land transfers and limitations on Kamiauti agreements failed to a large extent. It was well known then that legislation alone would not produce the desired changes. It was held for example that the Kamiauti system would only disappear with an improved economic situation to be partly brought about by strict enforcement of the tenancy law of 1908.[125] As far as the transfer of land was concerned, a problem which is intimately related to the question of supply of capital and credit-worthiness, it was hoped that the cooperative movement would take roots in Chota Nagpur[126] which, after 1908, had enjoyed a headstart with the Catholics of Ranchi under the guidance of the Jesuit missionary J.

[124]The latest attempt to abolish bonded labour by law is the Government of India "Bonded Labour Abolition Act" of 1975.

[125]*Palamau Settlement Papers*: No. 7P, P.W. Murphy, Director Land Records and Surveys, Bihar and Orissa, to SBR, Bihar and Orissa.

[126]MacPherson/Hallett (1917), p. 165.

Hoffmann. The movement, however, did not spread as anticipated and the farmers in Chota Nagpur are still victimized by traditional forms of moneylending.

To return now to the first round of settlement operations in Chota Nagpur (1902-1925), they were in many respects a success. For the first time since the beginning of British rule, it became possible to know and not to guess what the agrarian situation was like. This, for example, helped considerably with the drafting of the Chota Nagpur Tenancy Act of 1908. Not only did the courts possess now a specially tailored law, but also written records of rights which showed the different rights and duties pertaining to land. Conflicts over rent enhancement decreased. This was however, attributed not to the rent law, but to a declining interest of landlords in their rent income, as a result of increasing returns from other sources like bakasht cultivation and forest.[127] The most important success, however, was the improvement of the economic situation of the rural population over the next quarter of a century.[128] As has been mentioned earlier, population density per square mile of net area sown decreased, which can only be explained by a faster expansion of arable land, mainly rice-land, as compared to population. This again led to increased overall rice production per capita, which, however, does not necessarily augment per capita availability of rice due to the uneven pattern of landholdings. It is clear that the combination of agrarian legislation plus settlement operations in the 1920s successfully removed the man-made obstacles to agricultural expansion described earlier. The raiyats of Chota Nagpur reacted, as could be expected, immediately to the new opportunities. For the first time they could clear new land, convert up-land into rice-land and improve irrigation facilities in comparable safety. They were less afraid than formerly that a rapacious landlord would succeed in stealing the fruits of their toil. The results of greater productivity became visible when the second round of settlement operations was started in Ranchi in 1927.

Conclusion

The basic problem in Chota Nagpur was and probably still is the clash between two cultures, two ethnic groups. This is a situation known the world over and is not limited to India. It is a question of bringing culturally and economically isolated and different populations into the mainstream of a national economy and culture without depriving them of their identity and means of economic survival. In Chota Nagpur this process of integration and of necessary adjustment did not work. In fact, the tribal population fought violently against the new order of things introduced and

[127] Taylor (1940), pp. 56, 58.
[128] Ibid., p. 91.

established by Sanskritized Rajas and Hindu immigrants. The British did not make life easier. Pax Britannica deprived the tribals of the chance to successfully respond in arms to oppression and substituted instead a foreign and partial idea of justice. Although the mere presence of British power in Chota Nagpur was a factor for change, because it impeded the free development of social forces, the intentional British policy was a policy to preserve the status quo as understood by Calcutta and Westminster. In assessing 182 years of colonial rule in Chota Nagpur, something can probably be found on both sides of the scale. One should see, however, that the British built a modern India only in so far as it was convenient and advantageous to them. The industrial base of British-India remained in Manchester and Birmingham and was not developed, for instance, in a region like Chota Nagpur which possessed all the necessary primary resources such as labour, coal and iron ore. Foreign rule in Chota Nagpur did not lead to modernization and impeded socio-economic change, it upheld for a long time a system which was detrimental to agricultural productivity. The effects are still evident.

3

Dietmar Rothermund

Tenancy Legislation for Chota Nagpur: The Emphasis on Executive Protection

TOWARDS the end of the nineteenth century the British rulers of India showed more and more concern for the protection of the peasantry which had become a tenantry under their legal dispensation. While in earlier years of liberal confidence they had trusted their economic doctrines which taught them not to interfere with the forces of the market and to uphold the freedom of contract, the mutiny of 1857 and later instances of peasant unrest had instilled a fear in their minds that their Indian empire might be swept away by a storm of peasant fury. Their solicitude for indebted peasants and oppressed tenants was guided by political motives of imperial self-preservation. Therefore their measures were not designed so as to protect everybody who might deserve it but only those whom they considered to be a source of potential political trouble.[1]

There were differences of opinion whether such protection should be

The following abbreviations have been used for the footnotes of this chapter: LD = Legislative Department, Government of India. LEGCO = Proceeding of the Legislative Council.

[1]For a detailed study of the history and political background of tenancy legislation see D. Rothermund, *Government, Landlord, and Peasant in India—Agrarian Relations under British Rule, 1865-1935*, Wiesbaden 1978; for the specific background of Chota Nagpur see Schwerin, *Von Armut zu Elend. Kolonialherrschaft and Agrarverfassung in Chota Nagpur, 1858-1908*. Wiesbaden 1977.

achieved through executive action or remedial legislation. The tendency to codify British-Indian law and to define the legal basis of executive action implied that protective interference by administrative officers would also need legislative support. Thus remedial legislation had to be sponsored even by those who would have preferred the old methods of unfettered executive discretion. But there was still scope for further debates between those who advocated procedural and enabling legislation which was aimed at facilitating executive action and those who wanted to change the substantive law and to rely on the civil courts rather than on the arbitrary benevolence of the revenue officer. These different schools of thought were identified with the administrative traditions of two provinces of British India which embodied contrasting experiences: Panjab and Bengal. Whereas in the Panjab in the good old "non-regulation" days the district officer did not even have to descend from his horse in order to dispense justice, giving a quick verdict on the spot and dashing off to settle the next case with equal speed and uncontested authority, the Bengal officer spent long hours in his court as a district magistrate following the due process of law.

In keeping with these traditions tenancy legislation in these two provinces was based on very different principles. The Bengal Tenancy Act of 1885 and the Panjab Tenancy Act of 1887, therefore, provide a study in contrast which is of immediate relevance for Chota Nagpur, because subsequently both Panjab and Bengal precedent converged in the Chota Nagpur Tenancy Act of 1908. This convergence was no accident, because Chota Nagpur belonged to the province of Bengal, however due to the special problems of its tribal population it was treated like a "non-regulation" province in which the executive played a predominant role just as in the Panjab. In this way the Chota Nagpur Tenancy Act which matured rather late and then emerged in a most comprehensive form became a kind of coping stone of the edifice of British Indian agrarian legislation. In order to understand the position of this coping stone we must examine the structure of that edifice and explain why Chota Nagpur was destined to occupy this particular place in it.

The main problems which tenancy legislation tried to solve may be summed up in the following points: (1) Protection against eviction, (2) Determination of a fair rent, (3) Prevention of the alienation of land to non-agriculturists. The Bengal Tenancy Act of 1885 was an attempt to tackle these problems in keeping with the Bengal tradition of entrusting the protection of titles in land as well as the determination of rent to the civil courts. The occupancy right of the tenant was protected by the accrual of prescriptive rights after twelve years of uncontested occupation. In practice this so-called 12-year rule had been applied retrospectively by revenue officers when deciding whether to recognize a tenant's occupancy right or not. It had been turned into a prospective measure by the

Bengal Rent Act of 1859, but in this way it proved to be an invitation to the landlord to shift his tenants so as to prevent the accrual of prescriptive rights. The Act of 1885 corrected this mistake by making such shifts ineffective and assuring the accrual of prescriptive rights.

The act further introduced the new category of a tenure-holder, defined as someone who held more than 100 bighas (ca. 30 acres). The tenants of such a tenure-holder could also acquire an occupancy status by prescriptive rights. The act also put a rather arbitrary stop to the enhancement of rents by providing for what amounted to an annual increase of 1% regardless of the fact whether the rent was high or low to begin with. This provision was introduced after several attempts at establishing either customary or economic standards had failed. The freedom of transfer of occupancy holdings was left to local custom, and no attempt was made to prohibit the alienation of land to non-agriculturists.[2] The Panjab Tenancy Act of 1887 was designed from a very different point of view. Prescriptive rights were carefully avoided and instead of this statutory recognition was supposed to protect the various classes of tenants defined in the act. The determination of rent and every other aspect of the protection of the tenantry was left to the revenue officer whose powers were increased by this act and a companion revenue act which codified executive practice as it prevailed in the Panjab.[3] No attempt was made as yet to restrict land alienation to non-agriculturists. This was left to the Panjab Land Alienation Act which was passed thirteen years later.[4] But in the meantime the Panjab precedent had made its impact on a revised tenancy act for the Central Provinces of 1898[5] which combined statutory tenancy with a restriction of land alienation, and it was this precedent of the Central Provinces which was available to those who tried to draft tenancy legislation for Chota Nagpur around the turn of the century when tribal unrest gave new urgency to such measures.

Legislation for Chota Nagpur was not written on a clean slate. In fact several half-hearted and partial measures had been adopted to cope with the special problems of this area beginning with the Chota Nagpur Tenures Act of 1869. This act was passed soon after the first Panjab Tenancy Act of 1868 and, therefore, the Panjab precedent was taken into

[2]For an analysis of the making of the Bengal Tenancy Act see D. Rothermund, "The Bengal Tenancy Act of 1885 and its Influence on Legislation of other Provinces", in *Bengal Past and Present, July-Dec. 1967*, also reprinted in D. Rothermund, *The Phases of Indian Nationalism and other Essays*, Bombay, 1970.

[3]For the various notes and drafts concerning the Panjab Tenancy Act of 1887 see LD, Oct. 1887, 126-163.

[4]D. Rothermund, "Freedom of Contract and the Problem of Land Alienation in British India" in *SOUTH ASIA*, 1972 (No. 2).

[5]For the notes and drafts concerning the Central Provinces Tenancy Act of 1898 see LD, Oct. 1898, 37-73.

account in Chota Nagpur even at that early stage of remedial legislation.[6] But there was an important difference between these two acts. The Panjab Act was passed in order to validate an earlier settlement which some officers had tried to upset in the course of a revision settlement; it then merely restored the initial recognition of various rights in land.[7] The Chota Nagpur Tenures Act on the other hand deliberately interfered with the Permanent Settlement of Bengal. Furthermore, while Act X of 1859, the first tenancy act of the Bengal Presidency, had never been extended to the Panjab nobody quite knew whether it was supposed to be in force in Chota Nagpur or not. In introducing the Act of 1869 the legislators adopted the point of view that it was not in force; they also did not adopt the principle of prospective prescription as they were dealing with the restoration of ancient rights and not with the acquisition of new ones. They even thought that these ancient rights were so clear and easy to ascertain that retrospective prescription was not really necessary. The Special Commissioner working under the Chota Nagpur Tenures Act was supposed to investigate the origin of every tenure without any limitation. The administrator in charge of the bill quoted Henry Maine's speech on the virtues of prescription in the Panjab Tenancy Act debate and said that prescription may be useful in the Panjab where custom was indistinct but not in Chota Nagpur where everybody in a village could identify bhuinhari land.[8]

After some further consultations this feeling of certitude vanished, the bill was amended and the Special Commissioner was no longer supposed to restore all ancient rights but was to be guided by a 20-year period of prescription.[9] The new legislation also rigorously excluded lawyers and the civil courts from any interference in the work of the Special Commissioner. Opinions in the Bengal Legislative Council were divided in this respect. Ashley Eden made fun of a certain class of officials "who preferred the dispensation of what was called patriarchal justice under a tree".[10] But one of this class of officials promptly defended the measure by pointing out that the people are "becoming the tools of designing Bengallees (sic)".[11] Thus the priority of the revenue authorities in dealing with this matter was vigorously upheld. As an afterthought the Select Committee also inserted a clause providing for the commutation of service into money rent, but this problem required further legislation at a later stage.[12]

[6]For the notes on the Chota Nagpur Tenures Act see LD, March 1869, 14-29.

[7]For the debate on the Panjab Tenancy Act of 1868 see LEGCO Governor General, 19 Oct. 1868.

[8]LEGCO Bengal, 28 Nov. 1868 (Mr. Dampier).

[9]LD, March 1868, 14-29.

[10]LEGCO Bengal, 28 Nov. 1868.

[11]Ibid. (Mr. Thompson).

[12]LD, March 1869, 14-29.

Actually the need for further legislation about the commutation of service rents was in a way induced by the Bhuinhar Survey, the survey of tenants' rights conducted under the Act of 1869, because by recording the services rendered by the bhuinhar peasants the survey gave official recognition to a custom which was of doubtful validity. Originally beth begari, as the work on the landlord's fields was called, was restricted to such peasants who had been specifically settled by the landlord on land called *beth keta* under the condition that they would cultivate his *majhas.* Only at a later stage when the landlords had extended their majhas and needed more labour services they had imposed this also on other peasants for whom it was often more convenient to put in a few days of labour rather than paying more rent. When even the bhuinhar peasants admitted that they rendered such service to the landlord they lent support to the presumption that this was a universal custom. Thus the Bhuinhar Survey instead of reducing this practice helped to spread it even more.[13]

The most important deficiency of the Act, however, was that it dealt only with bhuinhari and majhas land and excluded the recording of *rajhas* land as well as of pure *khuntkatti* villages. This was obviously done in order to limit the operations to the most pressing problem which was that of reconciling the claims of landlords and peasants to land which was supposed to be under their exclusive control. Rajhas was settled with peasants who had practically an occupancy status and had made no special complaints and the pure khuntkatti villages could not complain for the very reason that they were still intact. Had they been included in the survey at this stage much of the later trouble could have been forestalled.[14] As far as its restricted scope was concerned the Bhuinhar Survey conducted for more than a decade after the passing of the act stirred up more unrest than it settled. It had also the undesirable effect of making recorded bhuinhari land a valuable commodity as its transfer had not been prohibited.[15] So what was supposed to be a measure for the restoration of traditional peasant proprietors finally encouraged outside investors who were certainly not in need of special protection.

While the Bhuinhar Survey was still in progress the government took another legislative initiative by enacting a modified version of Act X of 1859 as Chota Nagpur Landlord and Tenant Procedure Act of 1879.[16] As the name of the Act indicated the main point of introducing it was to get an adequate legal foundation for rent suits, rent enhancement etc. The usual 12-year rule of prospective prescription was also imported

[13]LD, Oct. 1906, 19-21, (S.S. Day, Asst. Settl. Officer).
[14]LD, Sept. 1903, (E. Lister, Settl. Officer, Ranchi).
[15]LD, Oct. 1906, 19-21.
[16]Schwerin, op. cit.

into Chota Nagpur in this way although it was of no particular importance to this area where ryots had a customary occupancy rights as long as they paid the rent and rendered their services, which were the real crux of the problem. In addition to the Bengal precedent the Act contained some specific local features such as rather ineffective provisions for the commutation of services, and privileged rent rates for khuntkatti and bhuinhari land as well as favourable rates for newly reclaimed rice land (korkar) as regulated by custom. In contrast with Bengal proper where an amendment of Act X of 1859 had transferred jurisdiction in rent cases to the civil courts this Act of 1879 retained the exclusive jurisdiction of the revenue courts. Nevertheless, nobody was quite happy with this half-hearted law which was neither here nor there and stayed in suspended animation between Bengal precedent and Chota Nagpur practice.

After the passing of the Bengal Tenancy Act of 1885 efforts were made to extend this piece of legislative craftsmanship as soon as possible to Chota Nagpur.[17] The "No Rent" campaign of the frustrated leaders of the peasant movement of the late 1880s gave added urgency to this drive.[18] But in order to make Chota Nagpur fit for the Bengal act it was necessary to tackle the commutation of services as service rents were incompatible with it. Work on a draft bill for this purpose started in 1891. When Anthony Macdonnell as a member of the Government of India wrote a note on the process of draft legislation in 1894 he still firmly held the belief that this was only a flanking measure to facilitate the introduction of the Bengal Tenancy Act and admonished everybody concerned to deviate as little as possible from the spirit of the Bengal act. He was particularly annoyed at the fact that the bill he had to review needlessly accentuated the revenue officer and ousted the civil courts, and emphasized that this antagonism to the civil courts is entirely foreign to the spirit of the provincial constitution and its gradual development.[19] But by the time the final Commutation Bill was introduced into the Bengal Legislative Council in 1897 the Bengal Tenancy Act was no longer favoured by the Commissioner of Chota Nagpur, and the former commissioner who had recommended its introduction and was now in charge of the bill as a member of the Government of Bengal had to beat a hasty retreat which he could cover up by pointing to the impending amendment of that act which justified a policy of "wait and see".[20] The reasons which the new Commissioner of Chota Nagpur gave for his rejection of the Bengal Tenancy Act were that the "settled

[17] LD, Oct. 1897, 6-24 (Note by A. Macdonnell).
[18] E. Lister as quoted in footnote 14.
[19] A. Macdonnell as quoted in footnote 17.
[20] LEGCO Bengal, 21 Aug. 1897 (Mr. Grimley).

ryot" did not fit in with Chota Nagpur practice, that the permissiveness of the Bengal act with regard to the free transfer of holdings would encourage the moneylenders to buy up the peasants' land, and that the superior landlord's right of resumption was equally threatened by the free alienation of intermediary tenures.[21] The commissioner more or less echoed the opinions of the European manager of the Maharaja of Chota Nagpur's estate.[22] Since these were the days of the great land transfer alarm such views were bound to attract immediate attention. Some observers pointed out that such a policy would benefit the landlords most because they could either eject tenants for arrears of rent without having to resort to a sale of the holding or they could connive at a transfer by means of the device of "surrender" which gave the landlord a firm hold over the whole transaction and left him the lion's share of the profit.[23]

The Commutation Act was thus passed without the introduction of its companion, the Bengal Tenancy Act. But it was anyhow the stillborn child of a well-meaning but ignorant legislature. Commutation was not compulsory but optional, either landlord or tenant could apply for it. But it was provided that the amount added to the money rent in lieu of services should not be more than one fourth of the original rent. As the value of the services rendered was in many cases higher than the money rent the Act remained a dead letter. Once again the legislators had missed the mark after looking at their aim for a long time.[24]

The era of ineffective, piecemeal legislation was brought to a dramatic end by Birsa's rebellion of 1899.[25] At the same time the Central Provinces Tenancy Act of 1898 provided the legislators with an ideal blueprint for a comprehensive tenancy act for Chota Nagpur. It also so happened that John Woodburn who had rejected the Bengal precedent while he was Chief Commissioner of the Central Provinces was Lieutenant Governor of Bengal at the crucial time when Birsa's rebellion forced the Government of Bengal to do something about further legislation. Woodburn ordered first of all the preparation of a record of rights of the storm centre of peasant unrest and subsequent legislation to be based on the information obtained in this way.[26] The khuntkatti villages were the major focus of attention. The peasants of these villages were declared to be neither occupancy ryots nor tenure-holders but were placed in a category by themselves, the *Mundari Khuntkattidars*. Their tenure was

[21] LD, Oct, 1897, 6-24 (Commissioner Chota Nagpur to Govt. of Bengal, 3 Aug. 1897).

[22] Ibid. (Reference to Mr. Peppe).

[23] LD, Sept. 1903, 31 (J.C. Twidell, Dy. Com. Singhbhum, 31 March 1903).

[24] LD, Sept. 1903, 31.

[25] For a discussion of tribal unrest see D. Schwerin "The Control of Land and Labour in Chota Nagpur" in this volume.

[26] For a summary of Woodburn's initiative see LEGCO Bengal, 24 Feb. 1903 (Mr. MacPherson).

made inalienable the more so because being a specially privileged one it would have attracted moneylenders as a particularly good investment.[27] The Panjab Land Alienation Act which provided at least for transfer among members of the same agricultural tribe was not taken into consideration as a precedent for Chota Nagpur although it was explicitly recommended by an Indian government pleader of Hazaribagh who mentioned that it would solve the problem of a restriction of credit which might be too severe for the peasants.[28] The Central Provinces precedent prevailed over the Panjab precedent under the guidance of John Woodburn. He submitted the respective proposals to the Government of India and pleaded for very quick remedial legislation with regard only to the main points at issue leaving everything else to a later consolidating enactment. But the Government of India wanted an inclusion of provisions about rent and occupancy rights and this caused some delay.[29] The administrators also waited for the report of the settlement officer who conducted the record of rights operations in Ranchi District and who greatly contributed to the new definition of the Mundari Khuntkattidar.[30] The Act which was finally passed in 1903 also contained better provisions for the commutation of service into money rents. The unworkable rule of a maximum addition of one fourth of the original rent was given up and the rate of commutation was left to the judicial discretion of the revenue authorities.[31] The Act "accentuated the revenue officer" even more than its predecessors. The Government of India was particularly worried about the finality and conclusiveness attached to the record of rights, because other provincial governments had been explicitly warned to claim only presumptive but not conclusive proof for their record of rights. It was pointed out that an early opportunity should be taken to amend the Act in this respect.[32] But the Government of Bengal vigorously defended the conclusiveness of this record of rights as it would greatly unsettle the peasantry if it was known that the record could be set aside by the dreaded civil courts.[33] However, as it turned out later on the judges of the civil courts found some means to get around this provision by admitting the conclusiveness of the record with regard to possession but not as far as the title in land was concerned.[34] In this way they could open up the whole case inspite of the

[27]LD, Sept. 1903, 31.

[28]Ibid., (Letter of G.K. Gupta, 28 March 1903).

[29]LD, Sept. 1903, 31.

[30]E. Lister as quoted in footnote 14.

[31]LD, Sept. 1903, 31.

[32]Ibid., (Letter Govt. of India to Govt. of Bengal, 23 Sept. 1903).

[33]Govt. of India, Dept. of Revenue and Agriculture, Land Revenue, Jan. 1904, 26-27, (Letter Govt. of Bengal to Govt. of India, 24 Nov. 1903).

[34]LD, Oct. 1906, 19-21, (Com. Chota Nagpur to Govt. of Bengal, 21 Aug. 1905).

intentions of the legislators. But the civil courts did not only frustrate the work of the settlement officer after his record was completed, they stymied his operations even before he appeared on the scene by granting *ex parte* decrees to landlords who sued their tenants for fancy arrears of rent the moment they heard that the officer was coming.[35] All this required further legislation and the preparations for a new draft bill began almost immediately after the passing of the Act of 1903.

In addition to the old procedural feud between the courts and the revenue authorities there were also some substantive issues which had been forgotten in this Act of 1903. While khuntkatti land was made inalienable bhuinhari land was not and, therefore, investors and moneylenders concentrated on the latter even more eagerly than before.[36] There was also no adequate legal protection for the indigenous custom of reclaiming wasteland by preparing rice terraces. This operation, called korkar, was one of the main activities of the tribal peasantry and according to local custom they could extend their cultivation in this way without any special permission and they would also hold it on privileged rent. According to the cumbersome British law this was an improvement which required the landlord's consent. The peasants who were unaware of this were shocked when clever landlords pounced upon them and deprived them of the fruits of their labour.[37] In due course they stopped this kind of land reclamation altogether, and this was one of the reasons why the extension of cultivation could not respond to population pressure and people had to emigrate and make a living elsewhere.

The new round of legislation was aimed at mending these flaws of the existing law. The main point of departure for these renewed efforts was a joint note prepared by the Commissioner of Chota Nagpur and the settlement officer in charge of the record-of-rights operations in which all the above-mentioned problems were highlighted. This note was written by order of the Government of Bengal and submitted in 1905.[38] With this note the two officers also forwarded several letters of the missionaries whom they had consulted because of their intimate knowledge of the tribal people. Father Hoffmann, the Jesuit missionary, reported at length about the problems which British law and the working of the courts had meant for the tribal peasants. He highlighted the fact that no comprehensive act for Chota Nagpur had ever been drafted and that the courts had to rely on the Bengal precedent for everything that was not covered by the specific acts for this area. In this way the notion

[35]LD, Oct. 1906, 19-21.
[36]Ibid.
[37]Ibid.
[38]Ibid. (Joint Note of Commissioner Chota Nagpur, Slacke, and Settl. Officer, Lister).

that every inch of land belonged to the landlord had been introduced into Chota Nagpur where it did not fit the facts at all. This he considered to be the root of all the existing evils. It had encouraged the landlords to raise exaggerated claims. On the other hand they never gave rent receipts to the peasants and often charged them twice or sued them for arrears of rent. Actually many peasants had therefore stopped paying rents to the landlord until they were sued for arrears because this was the only way in which they could get their payments confirmed. This practice had then led to the wrong impression that the landlords were the aggrieved and harassed party which needed government help in the recovery of rents. Father Hoffmann also described the reasons for the general distrust of the civil courts among the tribal people. Most sub-judges and lawyers were landlords or tenure-holders themselves or related to people who had an interest in land in the area and the peasants were well aware of this fact. They rather trusted the British revenue officers who could afford to be impartial.[39] Hoffmann's letter was thus a plea for comprehensive legislation rather than piecemeal amendments. But it took some more time until the Government of Bengal also adopted this view and produced a large scale consolidating act to replace the various specific acts.

Eventually the Chota Nagpur Tenancy Act of 1908 managed to combine three distinct strands of legislative precedent, the specific legislation for tribal tenures, the Central Provinces precedent which had entered the Chota Nagpur tenancy law under John Woodburn's guidance, and the Bengal precedent which had prevailed in Chota Nagpur even with respect to legal provisions which were not enacted for this administrative division of the province. The reconciliation of the diametrically opposed principles of the Central Provinces Tenancy Act and the Bengal Tenancy Act was, indeed, a very difficult task and it would never have been undertaken except for the special constellation which existed in Chota Nagpur in terms of the nature of agrarian relations as well as with regard to the events which influenced the course of legislation.

When the joint note of the two officers from Chota Nagpur was taken into consideration by the Government of Bengal in 1905 there was no plan as yet to attempt this kind of large scale consolidation. Initially such specific problems as the protection of korkar land and the restriction on the transfer of bhuinhari land attracted more attention than the deeper problem of reconciling the various strands of legislation. The narrowly circumscribed approach to specific problems would have led to another act of the type for which Chota Nagpur was by now notorious, if the Lieutenant Governor, Andrew Fraser, had not taken a bold step. He dropped a bill which had been introduced into the Bengal

[39]LD, Oct. 1906, 19-21. (Letter of Father Hoffmann, 28 May 1905).

Legislative Council in 1907 and convened a conference in Ranchi which was attended by administrators as well as by representatives of the interested parties.[40] This conference even served as a mock legislature voting on the various clauses of draft legislation whereby the official majority saw to it that the government would have the upper hand. Nevertheless some bargaining between the administrators and the landlords went on at this conference. Two questions were of major importance to the landlords as usual, the enhancement of rent and the extension of land under their immediate control. As far as rent was concerned they were faced with the possibility of the application of the Bengal formula of two annas in the rupee maximum enhancement every fifteen years.[41] Prices had doubled or even trebled in Chota Nagpur over the last twenty years and therefore the government could hold out the promise to the landlords to deal with them more generously than the Bengal formula would permit, if the fixation of rent would be left to revenue officers under rules which the government could revise from time to time.[42] The landlords agreed to this dictatorship of the revenue authorities in order to escape from the consequences of the Bengal formula. With regard to the extension of their majhas or bakasht land the landlords were anxious that the government should not limit it to that which was recorded at the time of the Bhuinhar Survey according to the provisions of the draft bill. The government made a concession to them and dropped this limitation so as to accommodate those who had acquired such land after that survey.[43] Closely connected with this question was the intention of the government to enable peasants who had been deprived of their land to sue for its restitution. Originally the provisions of the draft bill were rather stiff in this respect: they extended the period of limitation to 12 years and removed such suits for restitution from the civil courts to the revenue courts.[44] Any peasant who had lost his land before or in the course of the agrarian disturbances and then emigrated could come back after many years to reclaim his land. This measure was rigorously defended by the majority of the administrators. But W. H. Vincent who was the Judicial Commissioner of Chota Nagpur and was later on in charge of the bill in the Bengal Legislative Council did not like the exclusion of the civil courts. He expected that there would be trouble about this in the Council and he said so at a meeting with the officers of the Government of India when the Bengal officers had to discuss this bill with them. But the Chief

[40]LD, April 1907, 1-6; LD, June 1908, 33-36; LD, Nov. 1908, 14-21.

[41]LD, Nov. 1908, 14-21.

[42]LEGCO Bengal, 19 Sept. 1908 (W.H. Vincent).

[43]LD, June 1908, 33-36 (Note by J. Kerr, 4 June 1908).

[44]LD, June 1908, 33-36 (Letter Govt. of Bengal to Govt. of India 30 March 1908; Govt. of India to Govt. of Bengal, 18 June 1908).

Secretary to the Government of Bengal said that the clause should be left in the bill as a concession could be made in the Council, if necessary.[45] And this is exactly what happened. The Select Committee then removed the provision which transferred suits for restitution of land to the revenue courts and reduced the period of limitation to three years instead of the 12 years suggested earlier.[46]

The officers of the Government of India also noticed the clause about the enhancement of rent according to rules made by the provincial government and found this provision rather unusual. The only precedent which could be found by them was a clause in the Oudh Rent Act of 1886 which empowered the government to change the rent restrictions mentioned in the Act by notification.[47] There was some opposition to this provision in the Bengal Legislative Council but one of the Indian members informed the Council of the Ranchi negotiations and that settled the issue.[48] W.H. Vincent supported the provision according to which government could make rules about the enhancement of rent with a remarkable argument when he spoke in the Council. He pointed out that rent cases would be tried under the act also by junior revenue officers who were quite inexperienced, therefore it was necessary for the government to prescribe rules in order to obtain at least some uniformity in these cases.[49] This was an unusual admission of a deficiency in the wisdom of revenue officers which was normally praised or at least not openly doubted. It was on the other hand a kind of special pleading for the extraordinary power which the government assumed in this way. Vincent's judicial mind was obviously somewhat ill at ease in this respect.

The Chota Nagpur Tenancy Act of 1908 which emerged from these debates was a unique document incorporating a blend of devices of tenancy protection which had matured in the course of half a century of legislative efforts. There was the "settled ryot" and prospective prescription, but there were also statutory tenants of the Central Provinces type. There were the civil courts of the Bengal tradition but there were also the powerful revenue authorities according to the North Indian style. Internal contradiction was avoided by entrusting the revenue authorities with the task of specific protection which had been defined in the series of earlier acts and leaving general agrarian relations to the provisions of the Bengal Tenancy Act which were re-enacted by this Act

[45]See Vincent's earlier remarks as Judicial Commissioner, Chota Nagpur, LD, Oct. 1906, 19-21; for the meeting with officers of the Govt. of India see LD, June 1908, 33-36.

[46]LD, Nov. 1908, 14-21.

[47]Ibid., (Note by J.O. Miller, 23 Oct. 1908).

[48]LEGCO Bengal, 19 Sept. 1908 (D.P. Sarbadhikari).

[49]Ibid., (W.H. Vincent).

with the omission of such sections which could not fit into the Chota Nagpur pattern.[50] The Act also included the famous Chapter X of the Bengal act (Chapter XII of this Act) which provided a basis for further survey and settlement operations. In the course of the next two decades these operations covered the whole of Chota Nagpur.[51] The record of rights prepared in the course of these settlement operations had presumptive proof only just like in other provinces. The finality and conclusiveness of the record of rights of the Mundari Khuntkatti area was not extended to other parts of Chota Nagpur, but the tenancy act had a clause which enabled the government to order a conclusive record of rights by notification for such areas for which such a measure would be required.[52]

No major changes were made in Chota Nagpur tenancy law in the course of the subsequent years and most of its provisions are still in force today after the State has taken over the land from the zamindars. The role of the revenue authorities has been enhanced in that way. Some minor amendments of the Act of 1908 were made in 1914, 1920, 1923 and 1929. The 1920 amendment introduced some relaxation in the restriction of transfer. Land could now be transferred among members of the same communities somewhat along the lines of the Panjab Land Alienation Act which had earlier been neglected as a precedent for Chota Nagpur.[53] A rather controversial matter was the amendment of 1929 which circumvented the restriction of transfer in a specific way so as to enable coal mining companies to acquire land from the peasants. This had become a major problem because landlords gave leases to these companies but could not legally make over the land occupied by peasants. Illegal means were used to force the peasant to abandon his land. If land was spoiled due to subsidence no proper compensation was paid. The settlement officer who had prepared the report on the district of Manbhum, where the coal industry was most prominent, had suggested that the restriction of transfer should be given up completely so as to enable the peasants to get the full value of their land.[54] This would have also enabled displaced peasants to buy land elsewhere in Chota Nagpur. But the government was unwilling to give up this restriction and, therefore, special provisions had to be introduced which enabled the revenue authorities to sanction transfers of land to the mining companies and even to compel unwilling peasants to quit their land and accept a price fixed by the Deputy Commissioner. This was a rather extraordinary clause and did not fit well into a tenancy act as was pointed out by members of

[50]For a discussion of this adaptation see D. Schwerin, op. cit. p. 454ff.

[51]Ibid., p. 485f. for the progress of survey and settlement in Chota Nagpur.

[52]LD, Nov. 1908, 14-21.

[53]Schwerin, op. cit., p. 486.

[54]B.K. Gokhale, *Settlement Report Manbhum.*

the provincial legislative council.[55] In fact, it would have been more appropriate to amend the Land Acquisition Act for this purpose, but as that was a central act which the Government of India would not amend just in order to meet a local problem, the provincial government was compelled to put these provisions into the tenancy act.[56] The powers of the revenue authorities which were already very extensive were thus increased even more. They have remained with them even after independence and have put some strain on the integrity of the officers who were in a position to satisfy the wishes of important interests or to withhold their assent.

A further measure which throws some interesting light on the development of agrarian relations in Chota Nagpur was the Chota Nagpur Tenure-holders' Rent Account Act of 1929. It was enacted against the wishes of the British administrators and emerged from a private member's bill of the astute politician, K. B. Sahay, who represented Hazaribagh in the Legislative Council of Bihar and Orissa. He had tried several times to push his bill through the Council and finally got a majority for it which forced the government to go along with him. The problem which he had in mind was the liability of co-sharers for the payment of the rent to the superior landlord. Obviously the families of tenure-holders who had managed to acquire interests in land in Chota Nagpur in the previous decades had proliferated by now to such an extent that shares were in many hands and defaulting co-sharers could greatly embarrass those who paid their share of the rent but would nevertheless forfeit their land if the landlord sold up the whole tenure for arrears of rent. K.B. Sahay's bill was, therefore, of great importance to the rent receiving middle class which appropriated a good deal of the agricultural surplus of Chota Nagpur. In keeping with the style of tenancy legislation K.B. Sahay wanted the revenue authorities to do the business of these people. The Deputy Commissioner was supposed to compel the landlord to maintain separate rent accounts for such co-sharers. The British administrators dreaded this additional task and objected to the bill. When they could not help it any longer they saw to it that the measure would at least be restricted to fractional shares of undivided tenures, because additional surveys would have been required in order to establish separate accounts for divided tenures where individual co-sharers had claims to a part of a village here and a few acres there and so forth.[57] The fact that this act was placed on the statute book showed the political influence of the middle class which was keen to protect its interests by appropriate legislation. But the central issue of this act also highlighted a peculiar

[55] LEGCO Bihar and Orissa, 20 Jan. 1928 (D.P. Sinha).
[56] Ibid. (Mr. Sifton); see also LD, File 38/II/1929 Public.
[57] LD, File 64/II/1929 Public.

problem which was posed by the control of land by this class: the tenureholders were the effective landlords who exploited the peasantry more directly than the large zamindars, but though they were the effective landlords they were not necessarily very efficient because their interest in land was often subdivided among quarrelling heirs who wasted their time and energy in checking rent accounts instead of thinking of the improvement of agriculture. In calling upon the government to help them to do their business and to assure them of a regular and uncontested rental income the tenureholders revealed their own incompetence. Thus they paved the way for the eventual abolition of their rights in land. A government which is supposed to collect revenue and rent on behalf of intermediaries may as well do away with them altogether. But the political considerations which led the British to go in for tenancy protection also prevented them from taking radical steps against such intermediaries. They wanted to maintain the balance of power in the countryside. Only an Indian Government could try to achieve a new balance after independence. But when this was done the steps taken were still far from radical. The rights of the intermediaries were abolished with due compensation but the Chota Nagpur Tenancy Act remained the basic framework of agrarian relations and with it executive discretion survived as the arbiter of the peasant's fate.

4

D.C. Wadhwa

Zamindaris at Work (1793-1956): Case Studies in the Dhanbad District*

PERHAPS, no question connected with Indian agriculture has been discussed more frequently than that of the agrarian structure or the institutional framework of agricultural production which includes, besides other things, land-tenure, legal or customary, under which the land is owned, and land-tenancy, the system under which the land is operated and its produce divided between the operator and the owner. Whatever different shapes the land-tenures and tenancies might have taken in the crucibles of different cultures, they had a vital bearing on the efficient utilisation of agricultural land. At the time of the attainment of independence in 1947, raiyatwari (ryotwary) and zamindari (zemindary) were the two principal types of land-systems prevalent in different parts of the country.

Zamindari System

Unlike the raiyatwari system under which the primary rights of proprietorship, with unrestricted rights of transfer, are conferred upon the individual raiyats (ryots) who pay land-revenue, fixed by the Government not by contract but as a settlement, for their holdings directly to the Government; under the zamindari system (Permanent Settlement) the primary rights of proprietorship were conferred upon the zamindars (zemindars) who paid a certain fixed amount of land-revenue for their zamindaris (permanently-settled estates), in accordance with the terms of the Permanent Settlement, to the Government and collected rent from the tenants according to the terms of the contract. If that land-revenue fell into arrears, the estate was put to auction and sold to the highest bidder. The purchaser of that estate acquired the estate free of all encumbrances created since the time of the Permanent Settlement and obtained a statutory title.[1] The Dhanbad district, as indeed the whole of the State of Bihar, was under the zamindari system introduced in 1793 in Bengal, Bihar and Orissa.

Zamindaris in Dhanbad District

The district of Dhanbad was divided into a small number of big estates. There were seven zamindars in the district who among themselves held all the land of the district except that of the present Chandankiari and Chas *anchals* (next administrative units to the district). The latter belonged to the former estate of Panchet, major portion of which now lies in the Purulia district of West Bengal. The names of those seven estates are: Jainagar, Jharia, Katras, Nawagarh, Nagarkiari, Pandra and Tundi.

Though it has not been possible so far to locate the primary sources for tracing the origin of those estates, it is learnt from the secondary sources that they were there even at the time of commencement of the British rule. Some of them, it is said, were obtained by clearing the jungles, some by ejectment of the former possessors and some by purchase. It is further said that the estates of Jharia, Katras and Nawagarh were offshoots of the Palganj (now in the Hazaribagh district of Bihar) estate, and the Nagarkiari and Pandra estates were offshoots of the Tundi estate. It appears that, prior to the advent of the British, they acknowledged the suzerainty of the Panchet zamindar. Perhaps, only the Pandra estate paid a regular land-revenue direct to the Mohammedan rulers. All those zamindars claim to be Rajputs. The succession to all those estates was regulated by the law of lineal primogeniture and as such the estates always remained undivided and were not divisible amongst the sons of the last holder. The eldest son succeeded to the estate while the other sons got maintenance grants, either in cash or in kind, enjoyable during their life time. In most cases the grants were resumable on the death of the grantors and the grantees. Those grants of land by way of maintenance were, having regard to the real character of the impartible estates, deemed not as transfers of ownership, but as assignments of only the rents and profits. The lands that were subject to the grants were not absolutely severed from the estates, there being reversion

[1]Bengal Regulations 1 and 8 of 1793, 3 of 1794, 5 of 1795 and 2 and 27 of 1803; and Madras Regulations 25 of 1802, 2 of 1806 and 4 of 1822.

in favour of the zamindar. After the resumption of the grants, they formed part and parcel of the impartible estates and thus the estates remained undivided and in their entirety. The only exceptions to that rule were the estates of Nawagarh and Pandra. The Nawagarh estate was divided into four *kismats* (shares) before the Permanent Settlement. The Pandra estate was divided into three shares in 1795 as a result of the judgement in the Suit No. 5225, dated November 26, 1793, in the District Court of Birbhum. The partition was confirmed by the *Sadar Diwani Adalat* (chief civil court) on September 4, 1801. The third share was again divided into two equal parts in 1825. But the Agent to the Governor-General, after holding an investigation into the custom of succession in the estate, found the decree of the civil court dividing the estate as erroneous (*vide* Proceedings dated December 9, 1840).[2] Whatever had been done was not undone, but the rule of succession by primogeniture was strictly followed thereafter in all the shares of the estate. The territories of the Nawagarh estate, though divided into four shares, were, in fact, under the control of three zamindars, while those of the Pandra estate, though divided into four shares, were under the control of two zamindars, each having two shares. The whole of the Jainagar estate, which was sold in execution of a civil court decree in 1866, was bought by the Court of Wards on behalf of the Jharia zamindar.

Socio-Economic Conditions in Zamindari Areas

Much has been written, by way of criticism, about the socio-economic conditions prevailing in the zamindari areas in the country. The most obvious criticism, as pointed out by a Select Committee of the House of Commons, has been that "while the amount of revenue payable by the zamindars to the Government became fixed", no measures were taken "to define or limit the demand of the zamindar upon the raiyat".[3] The amount of rent paid by the tenants to the zamindars was, as mentioned by the Court of Directors, in their despatch to Bengal, "regulated neither by specific engagement, nor by the established rate . . . but by the arbitrary will of the zamindars",[4] a fact later confirmed by the findings of the

[2]Final Report and Return of the Pandra (*kismats* 1 and 3) Encumbered Estate enclosed with Letter No. 783, dated July 11, 1927, from the General Manager, Wards and Encumbered Estates, Manbhum, to the Deputy Commissioner, Manbhum (*vide* File No. 20-557 of 1927 of the Board of Revenue, Bihar and Orissa); and Letter No. 1642, dated November 25, 1935, from the General Manager, Wards and Encumbered Estates, Manbhum, to the Deputy Commissioner, Manbhum (*vide* File No. 20-271 of 1936 of the Board of Revenue, Bihar).

[3]*Report from the Select Committee of the House of Commons on the Affairs of the East India Company*—1831-1832, London, 1832, Vol. 1, p. 63.

[4]Despatches to Bengal, 15th January to 25th February, 1819, pp. 83-84.

Calcutta Civil Finance Committee when they wrote that "in the permanently settled districts nothing is settled, and little is known but the Government assessment".[5] As regards the amount of revenue paid by the zamindars to the Government in comparison to the rental collected by them, it was said that with the rise in value and produce of land, the assessment of the zamindars became very light and "tended to become", as observed by the Taxation Enquiry Committee, "more and more unequal as time went on"[6] with the result that "while the zamindars gained in strength, the cultivator", as mentioned in a Government Resolution, "was rack-rented, impoverished and oppressed".[7] The Indian Statutory Commission drew attention to sub-infeudation, encouraged by this system, which was considered an incubus on the working agricultural population.[8] The Land-Revenue Commission, appointed in 1938 to examine the then existing land-revenue system of Bengal in its various aspects with special reference to the Permanent Settlement, after pointing out its various defects of financial loss to the Government, absence of contact with the cultivators, absence of agricultural improvement, absence of remission of rent, problem of sub-infeudation, increasing loss of occupancy-rights, etc., concluded that "whatever may have been the justification for the Permanent Settlement in 1793, it is no longer suited to the conditions of the present time".[9]

Abolition of Zamindari System

As intermediaries were considered to be the main obstacle to progress in agriculture, the institution of zamindari was abolished in the whole of the country, wherever it existed, after the attainment of independence, with a view to creating peasant-proprietorship. In Bihar, the Bihar Legislature passed the Bihar Abolition of Zamindaris Bill, 1948 (enacted as Bihar Act 18 of 1949), to deprive the zamindars and tenure-holders of their estates and tenures respectively. This Act was, however, repealed

[5] *Report of the Calcutta Civil Finance Committee* submitted on 12th July, 1830—Bengal Financial Consultations, 6th July to 31st August, 1830, Consultation 3, dated 20th August, 1830, p. 2. In the *Report of Land-Revenue Commission, Bengal*, Calcutta, 1940, Vol. 1, at p. 23, this statement of the Calcutta Civil Finance Committee has been wrongly attributed to the Select Committee of the House of Commons, 1830.

[6] *Report of the Indian Taxation Enquiry Committee*—1924-1925, Madras, 1926, Vol. 1, p. 42.

[7] Resolution No. 1, dated 16th January, 1902, issued by the Governor-General-in-Council, published in the *Land-Revenue Policy of the Indian Government*, Calcutta, 1902, pp. 7-8.

[8] *Report of the Indian Statutory Commission*, London, 1930, Vol. 1, p. 340.

[9] *Report of the Land-Revenue Commission, Bengal*, Calcutta, 1940, Vol. 1. pp. 35-42.

by the Bihar Act 9 of 1950[10] and no action was taken under any of its provisions. But subsequently, the Bihar Land Reforms Bill, 1950 (enacted as Bihar Act 30 of 1950), was passed which practically incorporated all the provisions of the Bihar Abolition of Zamindaris Act, 1948 (Bihar Act 18 of 1949), with a few more additions. The zamindari-abolition scheme could not, however, be implemented at the time on account of firstly injunctions issued by the law-courts and then the declaration of this Act as invalid by the Patna High Court (*vide Kameshwar Singh* v. *The State of Bihar*, A.I.R (38) 1951 Patna 91 or I.L.R (30) 1951 Patna 454)*. The State Government preferred an appeal in the Supreme Court against the decision of the Patna High Court. But in the meantime, the Central Government with a view to putting an end to all this litigation[11] and to

*This citation stands for the '*All India Reporter*, Volume 38, 1951, Patna Section, page 91, or the *Indian Law Reports*, Volume 30, 1951, Patna Series, page 454.'

[10]In the *Bihar District Gazetteers*, *Dhanbad*, by P. C. Roy Chaudhury, Patna, 1964, at p. 731, it is wrongly mentioned that the Bihar Abolition of Zamindaris Act, 1948 (Bihar Act 18 of 1949), was declared invalid. The brief history of the Act is as follows: The Bihar Abolition of Zamindaris Bill, 1948 (enacted as Bihar Act 18 of 1949), after it was passed by the Bihar Legislature, was reserved for consideration of the Governor-General. The Bill was, however, returned to the Governor of Bihar by the Governor-General for reconsideration by the Bihar Legislature with certain suggested amendments. The Bill was passed again with amendments suggested by the Governor-General and received his assent on July 6, 1949, and was notified as the Bihar Abolition of Zamindaris Act, 1948 (Bihar Act 18 of 1949). A suit was instituted in a court of Darbhanga, another district of Bihar, for a declaration that the Act was *ultra vires* of the Bihar Legislature and for injunction prohibiting the Government of Bihar from taking any action under the said Act. The suit was in due course transferred to the High Court in Patna in its extraordinary original civil jurisdiction. While this suit was pending, another Bill entitled the Bihar Land Reforms Bill, 1950 (enacted as Bihar Act 30 of 1950), was introduced in the Bihar Legislative Assembly practically incorporating all the provisions of the Bihar Abolition of Zamindaris Act, 1948 (Bihar Act 18 of 1949), with a few more additions. By one of its clauses, the Bihar Government also sought to repeal the Bihar Abolition of Zamindaris Act, 1948 (Bihar Act 18 of 1949), but before the Land Reforms Bill, 1950, could be considered by the Bihar Legislature, the suit in respect of the Bihar Abolition of Zamindaris Act, 1948 (Bihar Act 18 of 1949), came up for hearing and the Patna High Court fixed 19th January, 1950, as the hearing date. A few days before the date of hearing of the suit, the Government of Bihar introduced another Bill entitled the Bihar Abolition of Zamindaris Repealing Bill, 1949 (enacted as Bihar Act 9 of 1950), which was passed by the Bihar Legislative Assembly and assented to by the Governor only a day before the date fixed for hearing of the suit. When the suit was taken up for hearing, the matter was placed before the court and after hearing the parties the court dismissed the suit on the ground that the cause of action had abated on account of the Bihar Abolition of Zamindaris Act, 1948 (Bihar Act 18 of 1949), having been repealed (*vide Kameshwar Singh* v. *The Province of Bihar*, A.I.R (38) 1951 Patna 246, at page 252 or I.L.R (29) 1950 Patna 252). Thus the Bihar Abolition of Zamindaris Act, 1948 (Bihar Act 18 of 1949), was never declared invalid.

[11]The states of Madhya Pradesh and Uttar Pradesh also had passed Acts for the

remove certain defects in the working of the Constitution and also to remove the legal obstacles in implementing the zamindari-abolition Acts already passed by different states, brought forward a bill to amend the Constitution which was passed as the Constitution (First Amendment) Act, 1951. This amendment saved the agrarian-reform legislation from being challenged in the law-courts by a twofold provision by inserting Articles 31A (*vide* section 4) and 31B (*vide* section 5) in Part III of the Constitution dealing with Fundamental Rights. Article 31A provided that no law (past or future) providing for the acquisition by the State of any estate [including *jagir*, *inam* or *muafi* (all revenue-free grants of land) or other similar grant] would be deemed to be void on the ground that it was inconsistent with, or took away or abridged any of the fundamental rights enumerated in that Part. Notwithstanding the general provisions made in Article 31A, Article 31B was added to protect the particular Acts, relating to such matters, passed by different states till then. They were specified in a new Schedule, namely the Ninth Schedule, added to the Constitution (*vide* section 14) and their validity was ensured by this Article which provided that without prejudice to the provisions contained in Article 31A, none of the enactments specified in that Schedule would be deemed to be void, or ever to have become void, on the ground that such enactment was inconsistent with, or took away or abridged any of the fundamental rights conferred by any provisions of that Part. It further provided that all enactments included in the Ninth Schedule would continue to be in force, in spite of any judgement to the contrary, till repealed or amended by a competent legislature. The Bihar Land Reforms Act, 1950 (Bihar Act 30 of 1950), topped the list of thirteen enactments included in the Ninth Schedule. As the insertion of these Articles was made retrospective by providing that they "shall be deemed always to have been inserted", the Bihar Land Reforms Act, 1950, stood validated *ab initio*, irrespective of the above-mentioned decision of the Patna High Court. The validity of this amendment to the Constitution was also challenged in the Supreme Court but it was declared valid (*vide S.P. Singh Deo* v. *The Union of India and the State of Bihar*, A.I.R (38) 1951 S.C. 458 or I.L.R (30)1951 Patna 1176)*. The

abolition of intermediary rights in those states. The landlords there also had attacked the validity of those Acts in the law-courts on the ground that they were *ultra vires* of the Constitution as they contravened the fundamental rights conferred upon them under Part III of the Constitution. Though the Patna High Court had held the Bihar Land Reforms Act, 1950 (Bihar Act 30 of 1950), as unconstitutional, the High Courts at Nagpur and Allahabad had upheld the validity of corresponding legislation in Madhya Pradesh and Uttar Pradesh respectively. Appeals against all those decisions were instituted in the Supreme Court.

*This citation is to be read for the '*All India Reporter*, Volume 38, 1951, Supreme Court Section, page 458, or the *Indian Law Reports*, Volume 30, 1951, Patna Series, page 1176.'

Supreme Court also upheld the validity of the Bihar Land Reforms Act, 1950 (*vide The State of Bihar* v. *Kameshwar Singh*, A.I.R (39) 1952 S.C. 252 or I.L.R (31) 1952 Patna 565) and paved the way for its implementation. Though the original Bihar Land Reforms Act, 1950, had provided that the State may acquire the estates or tenures of zamindar or tenure-holder from time to time, with the enactment of the Bihar Land Reforms (Amendment) Act, 1953 (Bihar Act 20 of 1954), outright acquisition of all the intermediary interests in the State was rendered possible.[12] But the lack of adequate number of administrative personnel, experienced or trained in zamindari-management, was proving the main stumbling block in embarking on any scheme of total abolition of the zamindari system simultaneously throughout the State. In view of this the State Government decided to carry out the acquisition of estates and tenures according to their income in a phased programme. The estates and tenures having gross annual income exceeding Rs. 50,000 were to be taken over immediately in the first phase.[13] The estates and tenures having gross annual income between Rs. 10,000 and Rs. 50,000 were to be acquired in 1954 in the second phase.[14] All the remaining intermediary interests were to be acquired on January 1, 1956, in the last phase.[15] In the Dhanbad district, the Jharia, Nawagarh and Tundi estates were taken over during the first phase and the remaining estates and tenures were acquired in 1956 during the last phase.

Aims and Objects of the Study

The evils of zamindari system have been brought out, in their varied aspects, by a number of writers. The problem of land-tenures has, however, two aspects, the legal framework and the actual working. The former has received more attention than it merits; the latter less. As there does not exist at present a single case study of the working of an individual zamindari in India, it was decided to take up the study of the working of

[12]After section 3 of the said Act, a new section, namely 3A, was inserted which provided that "the State Government may, at any time, by notification, declare that the intermediary interests of all intermediaries in the whole of the State have passed to and become vested in the State".

[13]Memo No. EL-5-208/52-1910-L.R., dated May 20, 1952, from the Additional Secretary to the Government of Bihar, Revenue Department, to all the District Officers including the Additional Deputy Commissioner, Dhanbad.

[14]Letter No. 1330-E-VII-108/1953-L.R., dated March 1, 1954, from the Deputy Secretary to the Government of Bihar, Revenue Department (Land Reforms Section), to all the Collectors including the Additional Deputy Commissioner, Dhanbad.

[15]Memo No. E/VII-1-22/55-4383-L.R.T., dated August 18, 1955, from the Deputy Secretary to the Government of Bihar, Revenue Department (Land Reforms Section), to the Collectors of Bhagalpur, Muzaffarpur, Patna, Ranchi, Santal Parganas, Saran, Shahabad and Singhbhum including the Additional Deputy Commissioner, Dhanbad.

individual zamindaris of the Dhanbad district. The study aims at describing and analysing the working of all the zamindaris of the district from 1793 to their abolition. There are several aspects, for example, the origin of those zamindaris, their permanent settlement, management of land by the zamindars, sub-infeudation, agricultural improvements, incidence of rents, occupancy-rights of the tenants, coal-mine leases, financial position of the zamindaris, litigation, management of the zamindaris under the Chota Nagpur Encumbered Estates Act, 1876 (Central Act 6 of 1876), abolition of the zamindari system and the compensation to be paid to the zamindars and tenure-holders, etc., which are under study and will be brought out in several volumes. The data about almost all the above-mentioned aspects of the study have been collected from different sources, that is, revenue-records maintained at the village, *anchal* and district levels; records of the Boards of Revenue and archives of Bengal and Bihar states; records of the District Courts, High Courts and the Privy Council; records of the zamindars; and material available at the India Office Library and Records, etc.

Present Study

The present contribution in this book intends to give, in brief, the management of land in all the zamindaris of this district and the financial position of two zamindaris. The chapter about the management of land does not contain a full account of the distribution of land, under the control of the zamindars and tenure-holders, into different heads, but gives the distribution of land between the zamindars and tenure-holders and the management, by the zamindars and tenure-holders, of only (a) land under their personal cultivation, (b) their uncultivated land, and (c) *raiyati* (settled with the protected tenants) land under their control. Similarly, the chapter about the financial position of the two zamindaris does not give a detailed account of the receipts and disbursements of those zamindaris but gives only their total income and expenditure to assess the problem and the size of their indebtedness. The fifth chapter entitled 'Zamindars and Their Land' is devoted to the study of management of land and the sixth chapter entitled 'Zamindars in Debt' presents the financial position of the zamindaris. The detailed account of the distribution of land as well as the receipts and disbursements of the zamindaris would be given in an independent book to be published later.

5

D.C. Wadhwa

Zamindars and Their Land*

THE district of Dhanbad came into existence at the time of the states reorganisation on November 1, 1956, when the Manbhum district in Bihar was bifurcated into the districts of Dhanbad and Purulia. All the territories of the Purulia sub-division (excluding the territories of the Chandankiari, Chandil, Chas, Ichagarh and Patamda police-stations) of the Manbhum district were transferred to the State of West Bengal and constituted into the Purulia district there. Of the territories of the remaining five police-stations of the Purulia sub-division, the territories comprising the Chandankiari and Chas police-stations were attached to the newly formed district of Dhanbad which included in it, besides these territories, all the territories of the Dhanbad sub-division (sub-district) of the Manbhum district. The territories included in the Chandil, Ichagarh and Patamda police-stations were added to the district of Singhbhum in Bihar. For the purposes of revenue-administration, the Dhanbad district is divided into ten *anchals*, namely, Baghmara, Baliapur, Chandankiari, Chas, Dhanbad, Govindpur, Jharia, Nirsa, Topchanchi and Tundi. Each *anchal* is divided into a certain number of *halkas* (lower administrative units to the *anchals*). Each *halka* contains a certain number of villages.

Scope of the Study

The district of Dhanbad comprises 1,626 villages. As the present study concerns the zamindars of this district only, all the two hundred and seventy-eight villages of the Chandankiari and Chas *anchals* and eleven villages of the Nirsa *anchal*, belonging to the former estate of Panchet, major portion of which now lies in the district of Purulia (West Bengal), were excluded from this study. This chapter therefore presents data about 1,337 villages of the district. These 1,337 villages are grouped into eighty *halkas* of eight *anchals*.

The data about the management of their land by all the three zamindars of the Nawagarh estate have been processed separately. However, they have been combined into one unit for the purposes of analysis in the present study. Similarly, the data about the management of their land by both the zamindars of the Pandra estate also have been clubbed together into one unit. The data of both these estates would be given and analysed zamindar-wise in the book to be published later. Though the whole of the Jainagar estate was bought by the Jharia estate in 1866, the data about the management of land in both these estates have been presented here separately. In the book to be published later, the data of these estates would also be given and analysed as though belonging to one estate, namely Jharia.

Period of Reference

As a result of the survey and settlement operations during 1918-1925, a complete record-of-rights was prepared for each village for the first time in this district, which gives the position prevailing at that time. That record-of-rights is therefore the basis of this study. Though the record-of-rights was prepared at the time of the survey and settlement operations during 1918-1925, the information recorded therein relates to the period from 1793 to that date. Therefore the year 1793 is the base line of this study. Though the zamindari system was abolished in the whole of the district by 1956, it was decided to take the year 1970 for studying the post-zamindari abolition position so as to be able to study the full impact of the abolition of the zamindari system on the distribution of land, if any, within fifteen years of its abolition. But as it is intended to give in the present study the distribution of land between the zamindars and tenure-holders and the management of their land by them, the year by which all the intermediary rights stood abolished, that is 1956, has been taken as the last year for the purposes of this study. Thus the present study relates to these two periods, namely 1793 and 1956, with 1918-1925 as an intermediate stage. The account of distribution of land between 1956 and 1970 would be given in the book to be published later.

Sources of Data

The sources of data for the study of land distribution have been the villagewise records of all the villages included in this study. The record-of-rights prepared for each village after the conclusion of the survey and settlement operations in the district and made use of for this study consists of *Khatian* (containing record in respect of each separate holding or tenancy), *Khewat* (record of intermediary interests in a village), *Khasra* (list of plots in a village together with the name of the tenant, area of the plot and other details), *Register C* (containing villagewise details of all proprietary interests) and *Register D* (giving zamindari mutations). Owing to the prevalence of the Permanent Settlement in the State, the Government did not have any reliable and up-to-date land-record at the time of the abolition of the zamindari system. The record prepared at the time of the survey and settlement operations had become out of date. In order to have the then prevalent position of the land-holdings in the State, which was required for the purposes of zamindari-administration, the Government collected information by means of plot-to-plot enquiries (known as *field-bujharat*) and prepared a land-record, after the abolition of the zamindari system, depicting the position prevailing at the time of the abolition of the zamindari system. After completing the *bujharat* work, the necessary entries were made in registers called *Tenantwise Bujharat Register of Tenants' Holdings*, *Landlordwise Bujharat Register of Bakasht Lands*, *Landlordwise Bujharat Register of Ghairmazrua Malik Lands*, *Bujharat Register of Ghairmazrua Am Lands* and *Dispute List* (*Relating to Tenants' Khatian*) which contained information about *raiyati*, *abad-malik* (land under personal cultivation of the zamindars and tenure-holders), *ghairabad-malik* (uncultivated land in possession of the zamindars and tenure-holders), *ghairabad-sadharan* (uncultivated land the rights in which are vested in the village community) and disputed land respectively. Similarly, *bujharat* of *Khewat* was also undertaken and the entries of *Khewat Part I* (giving the interests of the zamindars of the revenue-paying estates), *Khewat Part II* (containing the interests of the zamindars of the revenue-free estates) and *Khewat Part III* (containing the interests of the tenure-holders and sub-tenure-holders) of the last survey and settlement were made in *Khewat Bujharat Forms III, IV and V* respectively. In the Dhanbad district, as indeed in the whole of the Chota Nagpur Division, where the number of the zamindars was rather small, the interests of the zamindars and tenure-holders had been recorded in one *Khewat*. The data about the position in 1956 were collected from these villagewise records.

Distribution of Villages

Table 5.1, which gives distribution of villages between the zamindars and tenure-holders in all the estates of the district as per settlement (1918-1925), shows that out of 1,337 villages in the district as a whole, one hundred and sixty-one entire villages (12.04 per cent) were under the direct control of the

TABLE 5.1

DISTRIBUTION OF VILLAGES BETWEEN THE ZAMINDARS AND TENURE-HOLDERS IN DIFFERENT ESTATES OF THE DHANBAD DISTRICT AS PER SETTLEMENT (1918-1925)

Name of the estate	*Wholly under the control of the zamindars*	*Wholly under the control of the tenure-holders*	*Partly under the control of the zamindars and partly under the control of the tenure-holders*	*Total*
(1)	(2)	(3)	(4)	(5)
Jainagar	0	92 (100.00)	0	92
Jharia	53 (19.49)	117 (43.01)	102 (37.50)	272
Katras	12 (8.16)	62 (42.18)	73 (49.66)	147
Nawagarh	15 (10.14)	75 (50.68)	58 (39.19)	148
Nagarkiari	32 (32.65)	35 (35.71)	31 (31.63)	98
Pandra	18 (6.34)	211 (74.30)	55 (19.37)	284
Tundi	31 (10.47)	179 (60.47)	86 (29.05)	296
District Total	161 (12.04)	771 (57.67)	405 (30.29)	1,337

Note: Figures in brackets in columns 2, 3 and 4 are the percentages to the total number of villages given in column 5.

zamindars where no intermediary rights were created; seven hundred and seventy-one villages (57.67 per cent) were under the control of only the tenure-holders to whom the zamindars had transferred their proprietary rights; and four hundred and five villages (30.29 per cent) were partly under the control of the zamindars and partly under the control of the tenure-holders, that is, proprietary rights over a part of the land in those villages were given away by the zamindars to the tenure-holders, while the remaining land was retained by the zamindars under their own proprietorship. The Jainagar estate had passed completely into the hands of the tenure-holders. As this whole estate was sold in execution of a civil court decree in 1866, it is inferred that it might have been heavily encumbered and therefore the zamindar of that estate might have given away all his villages on *mukarrari* tenures (permanent leases with unalterable quit-rent reserved) to get at a time a large amount of *salami* (an entrance fee levied from a new tenant at the beginning of his tenancy or a gratuity or offering on receiving a lease) as well as the rental. In all the other estates also the number of entire villages under the exclusive control of the tenure-holders was much more than that of those under the exclusive control of the zamindars themselves though the degree of sub-infeudation varied from estate to estate. In the Pandra estate, for example, the percentage of villages under the exclusive control of the tenure-holders to the total number of villages in that estate was 74.30, while in the Tundi estate it stood at 60.47 and in the Nawagarh estate at 50.68. The percentage of villages under the exclusive control of the zamindars in the district as a whole to the total number of villages in the district was as low as 12.04. Only in two estates, namely Jharia and Nagarkiari, the percentage of villages under the exclusive control of the zamindars to the total number of villages in those estates was higher than that of the district. In all other estates it was lower than the district percentage.

If villages partly under the proprietorship of the zamindars and partly under the proprietorship of the tenure-holders are taken into account, it will be seen from the same Table that the tenure-holders had their intermediary interests in 87.96 per cent of the villages in the district. In the Pandra estate the tenure-holders had their intermediary rights in 93.67 per cent of the villages of the estate, while in the Tundi estate the percentage of villages in which the intermediary rights were created, to the total number of villages in that estate, was 89.52. Among the estates having coal-bearing lands, namely Jharia, Katras, Nawagarh and Pandra, Katras was worst in this respect. The zamindar of this estate had created intermediary tenures in 91.84 per cent of its villages.

As the zamindars were indulging in extravagant habits, the finances of their estates, as will be seen from the study of financial position of two zamindaris in the next chapter, were reduced to such straits that they could not even meet their liabilities on account of taxes except with great difficulty by borrowing money. The loans were like halters round the necks of the zamindars and with the halters they could not survive long unless they parted with their zamindari rights permanently in some of the villages which brought them huge sums of money at a time. In all cases of financial difficulties that was the only course left to them and they unhesitatingly resorted to that with the result that most of their villages passed either wholly or in part into the hands of tenure-holders. It will be seen from the next chapter that in one estate alone, namely Tundi, one hundred and four entire villages (ninety-one villages by way of *mukarrari* grants and thirteen villages as usufructuary mortgages) had passed into the hands of tenure-holders within a short period of sixteen years, that is from 1914 to 1930, in consideration of *salami* amounting to lakhs of rupees in addition to rental.

Distribution of Land

As for the distribution of land between the zamindars and tenure-holders, Table 5.2 shows that 67.56 per cent of the total area of the district was under the exclusive control of the tenure-holders, and the zamindars retained their exclusive control over only one-third, that is 32.44 per cent, of its area. The entire area of the Jainagar estate was under the control of tenure-holders as all the villages of this estate had passed into their hands in entirety. Among the estates partly under the control of the zamindars and partly under the control of the tenure-holders, the Pandra estate had the highest percentage (84.97) of its land under the control of the tenure-holders, as in the case of number of its villages. The Tundi estate stood second with 75.90 per cent of its area under the control of the tenure-holders. The estates of Nawagarh, Jharia and Katras, having mainly coal-bearing lands, had 57.64, 55.35 and 49.80 per cent, respectively, of their total area under the control of the tenure-holders.

TABLE 5.2

DISTRIBUTION OF LAND BETWEEN THE ZAMINDARS AND TENURE-HOLDERS IN DIFFERENT ESTATES OF THE DHANBAD DISTRICT AS PER SETTLEMENT (1918-1925)

(in acres)

Name of the estate	*Under the control of the zamindars*	*Under the control of the tenure-holders*	*Total area*
(1)	(2)	(3)	(4)
Jainagar	0.00	19,536.18 (100.00)	19,536.18
Jharia	57,015.60 (44.65)	70,687.59 (55.35)	1,27,703.19
Katras	23,051.63 (50.20)	22,872.36 (49.80)	45,923.99
Nawagarh	23,803.90 (42.36)	32,384.15 (57.64)	56,188.05
Nagarkiari	16,873.79 (53.97)	14,390.33 (46.03)	31,264.12
Pandra	18,610.66 (15.03)	1,05,238.44 (84.97)	1,23,849.10
Tundi	23,501.42 (24.10)	74,007.29 (75.90)	97,508.71
District Total	1,62,857.02 (32.44)	3,39,116.34 (67.56)	5,01,973.36

Note: Figures in brackets in columns 2 and 3 are the percentages to the total area given in column 4.

This leads to the conclusion that though the land of the whole district was settled permanently with the zamindars, they retained their control on a small portion of it and created, for reasons already given, a variety of tenure-holders who exercised full control over the land given to them by the zamindars. The data collected about sub-infeudation, which are yet to be processed, show that the tenure-holders created, very often, sub-tenure-holders under them on almost the same conditions on which they had obtained their tenures from the zamindars. Again, the same data show that it was not uncommon for the sub-tenure-holders to create sub-sub-tenure-holders in the same way in which they held land from the tenure-holders with the result that there was created a chain of intermediaries between the actual cultivators and the zamindars. At the time of the survey and settlement operations during 1918-1925, there were 5,997 tenure-holders and sub-tenure-holders of various descriptions in the district, as given below[1]:

Name of the tenure	Number of the tenure-holders	Legal status of the tenures
(1)	(2)	(3)
1. *Brahmotar*	1,796	Rent-free, non-resumable, heritable, partible and transferable grants of land to the Brahmins
2. *Debotar* and *Shibotar*	514	Non-resumable, impartible and inalienable rent-free or bearing a small unalterable quit-rent grants for the worship of some idols
3. *Kharposh*	221	Resumable rent-free maintenance grants
4. *Lakhiraj*	136	A kind of *brahmotar* tenure
5. *Mukarrari*	1,011	Permanent, heritable, partible and transferable leases with unalterable quit-rent but heavy *salami* was realised at the time of settlement
6. *Dar-mukarrari*	118	Second degree subordinate tenure of *mukarrari* tenure
7. *Patni*	41	Non-resumable grants created in lieu of perpetual *kharposh* grants usually for the maintenance of illegitimate sons or concubines
8. *Dar-patni*	28	Tenures created by under-letting in the second degree, that is, tenure within tenure
9. *Mahatran*	123	Non-resumable, heritable, partible and alienable rent-free or bearing a quit-rent,

[1]Gokhale, B.K., *Final Report of the Survey and Settlement Operations in the District of Manbhum*, Patna, 1928, p. 50.

(1)	(2)	(3)
		not liable to enhancement, grants made to persons other than Brahmins in recognition of their secular services
10. *Maljagir*	30	Non-resumable, impartible and inalienable police tenures
11. *Madhyasatwa* (*Madhyastha*) and *Maurusi Madhyasatwa* (*Madhyastha*)	940	Non-resumable, transferable, heritable and partible intermediate tenures, the rent being liable to enhancement
12. *Jamaisatwa*	198	Same as *madhyasatwa* though in the case of *madhyasatwa* tenures the tenants were already on the land when the tenure was created, while in *jamaisatwa* there was generally some waste-land still available for reclamation
13. *Ijara*	428	*Ijaras* were of three kinds, namely, *chirasthai*, *meadi* and *be-meadi*. *Chirasthai ijara* was a non-resumable, heritable and transferable tenure with rent liable to alteration. A *meadi ijara* was generally for a short period of three to seven years, with unalterable rent, resumable on the expiry of the term but neither transferable nor partible within that period. *Ijardars* holding over after the expiry of the term in the original *meadi pattas* or holding without *pattas* were recorded as *be-meadi* and were liable to resumption after due notice at the end of the agricultural year
14. *Khuntkatti* and *Pradhani*	76	Heritable, with unalterable quit-rent, grants to village headmen
15. *Nayabadi* (*Nau-abadi*) and *Jalsasan*	146	Cultivating leases mainly originating from reclamatory leases. *Nayabadi* leases were settled on small quit-rent for five or seven years after the jungles were cleared, and after that period the area brought under cultivation was assessed at the ordinary rate. *Jalsasan* leases were similar to *nayabadi* ones but with an essential condition that the tenant was to construct a tank or a reservoir from which the land to be brought under cultivation or other land already cultivated could be irrigated. Usually the land on which the tank or reservoir was constructed and a few *bighas* (measure of land, varying in extent in different parts of India) of rice land below that were given either rent-free or on a

(1)	(2)	(3)
		small quit-rent. The remainder of the area was assessed to rent as in the case of *nayabadi* leases
16. Mortgages	.191	Usufructuary mortgages

The data about the compensation paid or to be paid to the ex-intermediaries in the district, collected from the *anchal* offices, show that the total number of intermediaries having some sort of intermediary interest in some of the land in the district and entitled to get compensation from the Government on account of the abolition of the intermediary rights is 7,949. It means that 1,952 (7,949 minus 5,997) more tenure-holders and sub-tenure-holders had come into existence in the district during the period between 1918-1925 and the abolition of intermediary rights between 1952 and 1956. Some of the new tenure-holders and sub-tenure-holders might have been created by the zamindars and tenure-holders, while the remaining tenure-holders and sub-tenure-holders might have acquired that status owing to partition of the partible tenures and sub-tenures during that period. Though the data about the actual number of intermediaries created at each layer, types of tenure-holders and sub-tenure-holders created by the zamindars and tenure-holders respectively, the extent of their jurisdiction, the reasons for which they were created and their rights and obligations, etc., have been collected, they have not been processed as yet. But it was noted during the collection of the above-mentioned data that there were sometimes as many as seven rent-receiving middlemen for the same piece of land. It was natural under those conditions that the landlords neglected their duties towards agriculture and displayed little interest in the improvement of conditions of the tenants. This will be amply borne out by the facts presented in the following section on agricultural improvements.

Agricultural Improvements

In Rules 217 and 220 of the Wards Manual it was expressly provided that at least three per cent of the rental (current demand) of solvent estates should ordinarily be devoted to agricultural improvements. The local officials in the Chota Nagpur Division were unanimous in accepting the principle that expenditure on the adequate maintenance of agricultural works existing when an estate was brought under Government management should be the first charge on the funds of an estate even if the result was likely to delay the payment of debts, and that such expenditure should not be charged against the ordinary three per cent allotment for improvements, which should be reserved for new works of improvement and their maintenance.[2] A study of the finances of the encumbered and wards estates of this district, however, shows that the Government officials never put this principle into practice with respect to estates which were taken under management of the Government under the provisions of the Chota Nagpur Encumbered Estates Act, 1876, Central Act 6 of 1876 (the provisions of which are described in the next chapter), or the Court of Wards Act, 1879 (Bengal Act 9 of 1879). Table 5.3 gives current demands (rent and cesses), collections, income from other sources and amounts spent on agricultural improvements in different estates during Government management under the provisions of the Encumbered Estates Act and the Court of Wards Act. It shows that in the Katras estate against the current demand of Rs. 1,07,127 of the estate during the period it was under management under the provisions of the Court of Wards Act from 1924 to 1926, Rs. 332-8-0* only were spent on agricultural improvements[3] which works out to be 0.31 per cent, that is one-tenth of the prescribed standard of three per cent. Again, during the fifteen years (from 1926 to 1941) of management of this estate under the protection of the Encumbered Estates Act, Rs. 6,333-0-9 were spent on agricultural improvements while the current demand during this period was Rs. 7,32,681.[4] So here also less than one per cent of the current demand was spent on agricultural improvements. The Nagarkiari estate was under management of the Government for twelve years (from June 8, 1910, to May 1, 1922) under the provisions of the Encumbered Estates Act. The only work of agricultural improvement undertaken in this estate during this period was the re-excavation of *khas* (for use by the zamindar only for his own benefit) tank on which Rs. 1,653-9-3 were spent. The current demand of the estate for the period was Rs. 1,25,058. Thus 1.32 per cent of the current demand was spent on agricultural improvements which benefited the zamindar only.[5] The Pandra estate (*kismats* 1

[2]Letter No. 138/IIE-5 R.T., dated March 24, 1917, from the Secretary to the Government of Bihar and Orissa, Revenue Department, to the Secretary, Board of Revenue, Bihar and Orissa (*vide* File No. 20-46 of 1917 of the Board of Revenue, Bihar and Orissa).

*The sum written as here is to be read as 332 rupees 8 annas and 0 pies. The annas and pies were the units of small currency before the introduction of the system of decimal coinage. An anna was a sixteenth part of a rupee while a pie was a twelfth part of an anna.

[3]Final Report and Return of the Katras Wards Estate enclosed with Letter No. 958, dated August 14, 1926, from the Deputy Collector Incharge, Wards and Encumbered Estates, Manbhum, to the Deputy Commissioner, Manbhum (*vide* File No. 20-481 of 1926 of the Board of Revenue, Bihar and Orissa).

[4]Final Report and Return of the Katras Encumbered Estate enclosed with Letter No. 516, dated April 3, 1944, from the General Manager, Wards and Encumbered Estates, Manbhum, to the Deputy Commissioner, Manbhum (*vide* File No. 20-619 of 1944 of the Board of Revenue, Bihar).

[5]Final Report and Return of the Nagarkiari Encumbered Estate enclosed with Letter No. 69, dated January 20, 1923, from the General Manager, Wards and

TABLE 5.3

CURRENT DEMANDS, COLLECTIONS, INCOMES FROM OTHER SOURCES AND AMOUNTS SPENT ON AGRICULTURAL IMPROVEMENTS IN DIFFERENT ESTATES DURING THEIR MANAGEMENT BY THE GOVERNMENT UNDER THE PROVISIONS OF THE ENCUMBERED ESTATES ACT AND THE COURT OF WARDS ACT

(in Rupees)

Name of the estate	*Period of management*	*Current demand (rent and cesses)*	*Collections (rent and cesses including arrears and interest)*	*Income from other sources*	*Total income from land (columns 4+5)*	*Amount spent on agricultural improvements*
(1)	(2)	(3)	(4)	(5)	(6)	(7)
Katras	1924-1926	1,07,127 (0.31)	77,719 (0.43)	58,918-5-¼	1,36,637-5-¼ (0.24)	332-8-0
Katras	1926-1941	7,32,681 (0.86)	7,32,586 (0.86)	14,58,695-4-8	21,91,281-4-8 (0.29)	6,333-0-9
Nagarkiari	1910-1922	1,25,058 (1.32)	1,19,848 (1.38)	2,714-4-0	1,22,562-4-0 (1.35)	1,653-9-3
Pandra (*kismats* 1 and 3)	1898-1926	11,07,423 (0.79)	10,84,054 (0.81)	2,57,518-10-4	13,41,572-10-4 (0.66)	8,842-2-4
Pandra (*kismats* 2 and 4)	1904-1934	14,86,745 (0.85)	13,31,122 (0.95)	2,84,746-8-1½	16,15,868-8-1½ (0.78)	12,606-9-6

Source: Final Reports and Returns of the Katras, Nagarkiari, Pandra (*kismats* 1 and 3) and Pandra (*kismats* 2 and 4) Wards and Encumbered Estates. The details about these reports and returns are given in foot-note numbers 3, 4, 5, 6 and 7.

Note: Figures in brackets in columns 3, 4 and 6 are the percentages, spent on agricultural improvements, to the current demands, collections and total income from land given in columns 3, 4 and 6 respectively.

and 3) was brought under management of the Government under the provisions of the Encumbered Estates Act on February 2, 1898, and was released on April 17, 1926. During the twenty-eight years of its management by the Government, Rs. 8,842-2-4 were spent by the estate on excavation and re-excavation of *khas* tanks of the estate. The current demand of the estate during this period was Rs. 11,07,423. Therefore the percentage of rental spent on agricultural improvements in this estate during that period was 0.79.[6] Here also the works of improvement were

Encumbered Estates, Manbhum, to the Deputy Commissioner, Manbhum (*vide* File No. 20-83 of 1923 of the Board of Revenue, Bihar and Orissa).

[6]Final Report and Return of the Pandra (*kismats* 1 and 3) Encumbered Estate enclosed with Letter No. 783, dated July 11, 1927, from the General Manager,

beneficial to the zamindar alone. The position was more or less the same in the other branch of the Pandra estate (*kismats* 2 and 4) which remained under Government management for thirty years (from July 31, 1904, to October 1, 1934) under the provisions of the Encumbered Estates Act. The current demand of the estate during that period was Rs. 14,86,745 out of which a sum of Rs. 12,606-9-6 only was spent on agricultural improvements[7] which amounted to 0.85 per cent of the rental. In the case of the Tundi estate, against an estimated annual income of Rs. 51,000, from the rent and cesses, in the scheme prepared in 1931, when it was under Government management under the provisions of the Encumbered Estates Act, Rs. 500 only were provided for the repair of *khas* tanks of the estate[8] which comes to less than one per cent of the annual rental. In the final scheme prepared in 1934-1935, this amount was reduced to Rs. 300[9] which works out to be 0.59 per cent of the annual rental. It is doubtful whether even this amount was spent. The only work to have ever been effected in this estate appears to be the construction of two granaries. One of them was completed in 1942 at Tundi at a cost of Rs. 1,100 and the other at Maharajganj at a cost of Rs. 800.[10] But none of the two granaries was of any benefit to the tenantry as they were intended for storing only the estate's collection of grain. Moreover, such works cannot be considered as works of improvement as, according to the rules, the works of improvement selected were to be those which answered to the definition of improvement given in section 76 of the Bengal Tenancy Act, 1885, Central Act 8 of 1885.[11] According to this definition, improvement meant any work which added to the value of the holding such as the construction of wells, tanks, water channels and other works for the storage, supply or distribution of water for the purposes of agriculture; the preparation of

---

Wards and Encumbered Estates, Manbhum, to the Deputy Commissioner, Manbhum (*vide* File No. 20-557 of 1927 of the Board of Revenue, Bihar and Orissa).

[7]Final Report and Return of the Pandra (*kismats* 2 and 4) Encumbered Estate enclosed with Letter No. 1642, dated November 25, 1935, from the General Manager, Wards and Encumbered Estates, Manbhum, to the Deputy Commissioner, Manbhum (*vide* File No. 20-271 of 1936 of the Board of Revenue, Bihar).

[8]Report of the General Manager, Wards and Encumbered Estates, Manbhum, enclosed with Letter No. 1452, dated November 14, 1930, from the Deputy Commissioner, Manbhum, to the Commissioner, Chota Nagpur Division (*vide* File No. 20-349 of 1931 of the Board of Revenue, Bihar and Orissa).

[9]Revised Final Scheme of the Tundi Encumbered Estate enclosed with Letter No. 2008, dated November 16, 1934, from the General Manager, Wards and Encumbered Estates, Manbhum, to the Deputy Commissioner, Manbhum (*vide* File No. 20-314 of 1935 of the Board of Revenue, Bihar and Orissa).

[10]Letter No. 1204, dated October 3, 1945, from the General Manager, Wards and Encumbered Estates, Manbhum, to the Deputy Commissioner, Manbhum (*vide* File No. 20-656 of 1945 of the Board of Revenue, Bihar).

[11]Correction Slip inserting Rule 219A in the Bihar and Orissa Manual. 1920 (*vide* File No. 20-245 of 1922 of the Board of Revenue, Bihar and Orissa).

land for irrigation; the drainage, reclamation from rivers or other waters, or protection from floods, or from erosion or other damage by water, of land used for agricultural purposes, or waste-land which was culturable; the reclamation, clearance, enclosure or permanent improvement of land for agricultural purposes, the renewal or reconstruction of any of the foregoing works, or alterations therein or additions thereto; and the erection of a dwelling house for the tenant and his family.

It can be argued that full demand was never realised in any estate at any time and therefore three per cent of the demand could not have been spent on agricultural improvements. But Table 5.3 shows that the expenditure on agricultural improvements in different estates under Government management was not only much below of the statutory obligation of three per cent of the rental (current demand of rent and cesses) but also much less than three per cent of the actual collections (rent and cesses including arrears and interest, if any). In fact, the position was almost the same in both the cases because there was not much difference between the collections and current demands.

The standard rate of three per cent, fixed by the Government, to be spent on agricultural improvements was only with respect to the current demand of an estate. But it will be seen from Table 5.3 that the estates had, in addition to the income from their current demands, sizable income from other sources also. The Katras estate, for example, had an income of Rs. 58,918-5-$\frac{1}{4}$, during Government management from 1924 to 1926 under the provisions of the Court of Wards Act, from the sale proceeds of trees, bamboos, mica and fish, *salami*, mutation fees, royalty on sand, stone and earth and excess royalty on coal. The same estate had an income of Rs. 2,70,960-4-11 from miscellaneous sources like *salami*, mutation and prospecting fees, sale proceeds of trees, bamboos, fruits, paddy, fish and coal, royalty on stone, sand, earth, bricks, gravels and fuel and excess royalty on coal, rent of fisheries, markets and fruit-gardens and the produce from forests, etc., during the fifteen years (from 1926 to 1941) of management of the estate by the Government under the provisions of the Encumbered Estates Act. If Rs. 11,87,734-15-9, the amount of royalty payable to the estate during this period and adjusted towards the repayment of loan borrowed by the estate from the Nawagarh Coal Company, are added to this income, the additional income of the estate during the period amounts to Rs. 14,58,695-4-8.[12] The current demand of the estate during that period was Rs. 7,32,681. It shows that the income from other sources of the estate was much more (almost double) than the current demand

[12]Final Report and Return of the Katras Encumbered Estate enclosed with Letter No. 516, dated April 3, 1944, from the General Manager, Wards and Encumbered Estates, Manbhum, to the Deputy Commissioner, Manbhum (*vide* File No. 20-619 of 1944 of the Board of Revenue, Bihar).

during the same period. The Nagarkiari estate had an additional income of Rs. 2,714-4-0 from *salami* on trees and the sale proceeds of trees, etc., during the period it was under Government management.[13] The income of the Pandra estate (*kismats* 1 and 3) from miscellaneous sources (*salami*, mutation and prospecting fees, sale proceeds of trees and paddy, royalty on coal, ballast stone and shooting and income from tusser) was Rs. 64,834-10-4 during the period it was under Government management. If Rs. 1,92,684, the dead rent, adjusted towards the assigned debts due to the East India Coal Company Limited, are added to it, the income of this estate from other sources amounts to Rs. 2,57,518-10-4.[14] Similarly, the other branch of the Pandra (*kismats* 2 and 4) estate realised an amount of Rs. 2,84,746-8-1½ from *salami*, mutation and prospecting fees, sale proceeds of paddy, trees and fish, royalty on coal, fireclay, ballast stone, sand, bricks and shooting and as income from tusser, etc., during the period of its management by the Government.[15] But these receipts of the estates from other sources were not taken into account while fixing the rate to be spent by them on agricultural improvements. If the other receipts of the zamindars, which accrued to them from the land, are also taken into consideration, in addition to their income from the rent and cesses (collections), the percentage of their total income from the land spent on agricultural improvements in the Katras (from 1924 to 1926), Katras (from 1926 to 1941), Nagarkiari, Pandra (*kismats* 1 and 3) and Pandra (*kismats* 2 and 4), estates comes to 0.24, 0.29, 1.35, 0.66 and 0.78 respectively.

The main argument, advanced by the officials incharge of management of the encumbered estates, for not spending on works of improvement in the district was that the zamindars had never maintained any irrigation works of real value and that consequently the acceptance of the principle that preference should be given to existing works of irrigation, when an estate was brought under management, was of no practical value. The encumbered estates could hardly be required to spend all that money on new schemes, they argued. There were, no doubt, in those estates, many areas in which there were few important works, but even where existing

[13]Final Report and Return of the Nagarkiari Encumbered Estate enclosed with Letter No. 69, dated January 20, 1923, from the General Manager, Wards and Encumbered Estates, Manbhum, to the Deputy Commissioner, Manbhum (*vide* File No. 20-83 of 1923 of the Board of Revenue, Bihar and Orissa).

[14]Final Report and Return of the Pandra (*kismats* 1 and 3) Encumbered Estate enclosed with Letter No. 783, dated July 11, 1927, from the General Manager, Wards and Encumbered Estates, Manbhum, to the Deputy Commissioner, Manbhum (*vide* File No. 20-557 of 1927 of the Board of Revenue, Bihar and Orissa).

[15]Final Report and Return of the Pandra (*kismats* 2 and 4) Encumbered Estate enclosed with Letter No. 1642, dated November 25, 1935, from the General Manager, Wards and Encumbered Estates, Manbhum, to the Deputy Commissioner, Manbhum (*vide* File No. 20-271 of 1936 of the Board of Revenue, Bihar).

works were both few and inconsiderable, as unfortunately was too often the case, there was all the more reason for the application of the principles accepted by the local officials.

Though under the Rules three per cent of the rental in solvent estates only was to be devoted to agricultural improvements, in 1938 the Government directed that at least three per cent of the rental should be spent on irrigation in all the estates under Government management irrespective of the debt obligations of the estates. In estates which were solvent, schemes for irrigation improvements were to be drawn up and amounts more than three per cent were to be spent for a few years to put the irrigation arrangements in order.[16] In continuation of this Order the Government further directed that in estates under Government management in which irrigation had been neglected for several years continuously, a greater amount than three per cent of the rental should be spent on irrigation in the beginning for a few years, according to the needs of each estate, as it was not sufficient to spend only three per cent of the rental on irrigation.[17]

As these orders were mandatory, the Government was requested by the Board of Revenue, at the instance of the Commissioner of the Chota Nagpur Division, to reconsider its orders with respect to the encumbered estates in Chota Nagpur as those estates there were heavily in debt and most of them failed to work up to the scheme arrangement for want of funds and therefore were not in a position to spend that much money on irrigation.[18] Consequently the Government agreed to the proposal that the Wards Department might regulate expenditure on irrigation in the encumbered estates according to the available income. Where funds were available every effort was to be made to improve the irrigation arrangements in accordance with the previous instructions of the Government. Where funds were not available those instructions were not to be observed until the financial condition of the estates had improved.[19] Though the matter was raised by the Commissioner of the Chota Nagpur Division only, the commissioners of all the Divisions in the State were informed about it as the orders of the Government seemed to be of general nature and affected

[16]Letter No. 6719 R., dated September 13, 1938, from the Assistant Secretary to the Government, Revenue Department, to the Secretary, Board of Revenue, Bihar (*vide* File No. 20-535 of 1938 of the Board of Revenue, Bihar).

[17]Letter No. 7883/IW-161/38 R., dated November 21, 1938, from the Secretary to the Government, Revenue Department, to the Secretary, Board of Revenue, Bihar (*vide* File No. 20-535 of 1938 of the Board of Revenue, Bihar).

[18]Letter No. 20-620-2, dated March 14, 1940, from the Secretary, Board of Revenue, Bihar, to the Secretary to the Government, Revenue Department (*vide* File No. 20-620 of 1940 of the Board of Revenue, Bihar).

[19]Letter No. W260/IW-27 R.R., dated May 29, 1940, from the Secretary to the Government, Revenue Department, to the Secretary, Board of Revenue, Bihar (*vide* File No. 20-620 of 1940 of the Board of Revenue, Bihar).

all the encumbered estates. There was nothing in the Chota Nagpur Encumbered Estates Act requiring the Manager of encumbered estates to give repayment of debts preference to expenditure on agricultural improvements. On the contrary, it was only after providing for such expenditure as Government dues, rent due to the superior landlords in the case of tenure-holders, sums requisite for the maintenance of holder of the property and his family, repayment of loans raised by the Government for meeting requirements of the estates and the costs of such repairs and improvements of the property as appeared necessary to the Manager, Wards and Encumbered Estates, and were approved by the Commissioner, the residue, if any, was to be applied to the discharge of the costs of management and the settlement of debts and liabilities of the estate (*vide* section 4 of the Chota Nagpur Encumbered Estates Act). However, little was done in the matter of agricultural improvements in such estates and the main concern of the Manager had always been the repayment of huge debts as has been observed from the study of finances of the encumbered estates. The *ahars* (reservoirs for collecting water for irrigation) and *bandhs* (dams) were allowed to fall into disrepair with the result that the condition of the tenantry became bad and the collection of rental fell. The depression of 1930's further reduced the ability of the tenants to pay their rents and aggravated the situation.

Table 5.4 gives information about the total demands (consisting of current demands of rent and cesses, arrears and interest), collections (rent and cesses including arrears and interest) and remissions in the Katras, Nagarkiari, Pandra (*kismats* 1 and 3) and Pandra (*kismats* 2 and 4) estates during the periods of their management by the Government. The Katras estate was brought four times under management of the Government under the provisions of the Court of Wards Acts and the Encumbered Estates Act. It has not been possible so far to trace the completion report and return and the audit report of the first period of its management from 1870 to 1884 when it was under Government management first under the provisions of the Court of Wards Act, 1870 (Bengal Act 4 of 1870), from 1870 to July 30, 1879, and then the Court of Wards Act, 1879 (Bengal Act 9 of 1879), from July 31, 1879, to 1884, after the former Act was repealed by the latter. During the second period of its management from 1896 to 1907 under the protection of the Encumbered Estates Act, the total demand of the estate was Rs. 3,17,161 out of which Rs. 2,81,938 (88.89 per cent) were collected. Out of the balance of Rs. 35,223, Rs. 7,779 (22.08 per cent) were remitted and the remaining amount was reported to be realisable.[20] As regards the collections during the third period of its management from 1924 to 1926 under the provisions of the Court of Wards Act, out of the total demand

[20]Report on Wards' and Attached Estates in the Lower Provinces, for the year 1907-1908, Calcutta, 1908, p. 28.

TABLE 5.4
TOTAL DEMANDS, COLLECTIONS AND REMISSIONS IN DIFFERENT ESTATES DURING THEIR MANAGEMENT BY THE GOVERNMENT UNDER THE PROVISIONS OF THE ENCUMBERED ESTATES ACT AND THE COURT OF WARDS ACT

(in Rupees)

Name of the estate	*Period of management*	*Total demands (current demands, arrears and interests)*	*Collections (rent and cesses including arrears and interest)*	*Balance*	*Remissions*
(1)	(2)	(3)	(4)	(5)	(6)
Katras	1896-1907	3,17,161	2,81,938 (88.89)	35,223	7,779 (22.08)
Katras	1924-1926	2,51,968	77,719 (30.84)	1,74,249	1,119 (0.64)
Katras	1926-1941	21,24,550	7,32,586 (34.48)	13,91,964	1,59,035 (11.43)
Nagarkiari	1910-1922	1,50,032	1,19,848 (79.88)	30,184	20,021 (66.33)
Pandra (*kismats* 1 and 3)	1898-1926	12,14,896	10,84,054 (89.23)	1,30,842	76,921 (58.79)
Pandra (*kismats* 2 and 4)	1904-1934	17,04,391	13,31,122 (78.10)	3,73,269	2,65,016 (71.00)

Source: Reports and Returns of the Katras, Nagarkiari, Pandra (*kismats* 1 and 3) and Pandra (*kismats* 2 and 4) Wards and Encumbered Estates. The details about these reports and returns are given in foot-note numbers 20, 21, 22, 23, 24 and 25.

Note: Figures in brackets in columns 4 and 6 are the percentages to the total demands and balance given in columns 3 and 5 respectively.

of Rs. 2,51,968 of the estate, only Rs. 77,719 (30.84 per cent) could be collected. Though out of the balance of Rs. 1,74,249, Rs. 69,827 (40.07 per cent) were officially reported bad, remission was granted for Rs. 1,119 (0.64 per cent) only.[21] The total demand of the estate during the last fifteen-year period (from 1926 to 1941) of its management by the Government under the provisions of the Encumbered Estates Act was Rs. 21,24,550 out of which a sum of Rs 7,32,586 (34.48 per cent) only was collected. Out of the balance of Rs. 13,91,964, Rs. 1,59,035 (11.43 per cent)

[21]Final Report and Return of the Katras Wards Estate enclosed with Letter No. 958, dated August 14, 1926, from the Deputy Collector Incharge, Wards and Encumbered Estates, Manbhum, to the Deputy Commissioner, Manbhum (*vide* File No. 20-481 of 1926 of the Board of Revenue, Bihar and Orissa).

were remitted.[22] This does not mean that the remaining sum of Rs. 12,32,929 (88.57 per cent of the balance) to be realised was realised. Though in the Pandra (*kismats* 1 and 3) estate 89.23 per cent of the total demand was realised during Government management from 1898 to 1926 under the protection of the Encumbered Estates Act,[23] in the Nagarkiari and Pandra (*kismats* 2 and 4) estates more than twenty per cent of the rental could not be realised during the periods they were under management of the Government. The total demand of the Nagarkiari estate during the period it was managed by the Government from 1910 to 1922 under the provisions of the Encumbered Estates Act was Rs. 1,50,032. Out of this Rs. 1,19,848 (79.88 per cent) were collected. Of the remaining Rs. 30,184, two-thirds, that is Rs. 20,021, were remitted.[24] In the Pandra (*kismats* 2 and 4) estate, Rs. 13,31,122 (78.10 per cent), out of the total demand of Rs. 17,04,391 during the period it was under Government management from 1904 to 1934 under the provisions of the Encumbered Estates Act, were collected and Rs. 3,73,269 remained to be realised of which Rs. 2,65,016 (71 per cent) were remitted.[25] It is reported that even these collections in all the estates were due to the excesses committed, in realisations, by the Government officials who were reported to be worse than the zamindars themselves in the matter of collections. Even when the tenants could not pay, remissions were not allowed except in very extreme cases.

If *ahars* and *bandhs* had been repaired, it would have meant gain to the tenants in the form of more produce and to the zamindars in the form of better realisations. Security of the rice crop depended largely on the extension of facilities for irrigation, and the tendency even among the landlords was to extend such cultivation beyond the limits of the area protected by its position or by such irrigation works as existed without adding to the number or capacity of the latter. Various suggestions seem

[22]Final Report and Return of the Katras Encumbered Estate enclosed with Letter No. 516, dated April 3, 1944, from the General Manager, Wards and Encumbered Estates, Manbhum, to the Deputy Commissioner, Manbhum (*vide* File No. 20-619 of 1944 of the Board of Revenue, Bihar).

[23]Final Report and Return of the Pandra (*kismats* 1 and 3) Encumbered Estate enclosed with Letter No. 783, dated July 11, 1927, from the General Manager, Wards and Encumbered Estates, Manbhum, to the Deputy Commissioner, Manbhum (*vide* File No. 20-557 of 1927 of the Board of Revenue, Bihar and Orissa).

[24]Final Report and Return of the Nagarkiari Encumbered Estate enclosed with Letter No. 69, dated January 20, 1923, from the General Manager, Wards and Encumbered Estates, Manbhum, to the Deputy Commissioner, Manbhum (*vide* File No. 20-83 of 1923 of the Board of Revenue, Bihar).

[25]Final Report and Return of the Pandra (*kismats* 2 and 4) Encumbered Estate enclosed with Letter No. 1642, dated November 25, 1935, from the General Manager, Wards and Encumbered Estates, Manbhum, to the Deputy Commissioner, Manbhum (*vide* File No. 20-271 of 1936 of the Board of Revenue, Bihar).

to have been made from time to time as to the steps to be taken to encourage landlords and tenants to make improvements. The most valuable form of convincing the tenants about the advantages of irrigation was to set a good example, but, far from doing so, all the estates in the district had continuously set the worst possible example by stinting expenditure on restoration and maintenance of existing works and by spending nothing on new improvements of this kind.

Land under Personal Cultivation

Under the provisions of section 6 of the Bihar Land Reforms Act, 1950, all the zamindars and tenure-holders were allowed to retain with them and hold as raiyats having occupancy rights under the State all lands used for agricultural or horticultural purposes which were in their own possession on the date of the abolition of the zamindari system, subject to the payment of such fair and equitable rent as was determined by the Deputy Commissioner in the prescribed manner. This also included (a) their privileged lands referred to in section 43 read with section 118 of the Chota Nagpur Tenancy Act, 1908 (Bengal Act 6 of 1908), that is, (i) lands which were cultivated by the zamindars and tenure-holders themselves with their own stock or by their own servants or by hired labour, or let out by them on leases for terms exceeding one year, or on leases, written or oral, for a period of one year or less, and which were, by custom, recognised as privileged lands in which occupancy-rights could not accrue, (ii) lands which were known as *man* [lands relinquished by the raiyats or purchased from them in rent-sales were called *man* as soon as they were reserved by the zamindars for personal cultivation in accordance with the provisions contained in sub-section (4) of section 72 of the above-mentioned Chota Nagpur Tenancy Act] in this district, and (iii) lands which were entered as *manjhihas* (lands reserved for the use of the zamindars of the villages, and at their absolute disposal) or *bethkheta* (lands ordinarily assigned as remuneration to the villagers who worked for the zamindars or their assigns on the *manjhihas* lands) in any register prepared and confirmed under section 5 of the Chota Nagpur Tenures Act, 1869 (Bengal Act 2 of 1869); (b) lands used for agricultural or horticultural purposes and held in the direct possession of their temporary lessees and cultivated by themselves with their own stock or by their own servants or by hired labour or with hired stock; and (c) lands used for agricultural or horticultural purposes and in possession of mortgagees, which immediately before the execution of mortgage deeds were in their own possession. With such a liberal allowance made for the retention with them of their lands under their personal cultivation, it was thought that they would be found to have large areas under their personal cultivation at the time of the abolition of the zamindari system. But Table 5.5, which gives information about the land under personal cultivation of the

TABLE 5.5
LAND UNDER PERSONAL CULTIVATION OF THE ZAMINDARS AND TENURE-HOLDERS AS PER SETTLEMENT (1918-1925) AND *BUJHARAT* (1956)

(in acres)

Name of the estate	*Zamindars*		*Tenure-holders*	
	As per Settlement	*As per Bujharat out of 'As per Settlement'*	*As per Settlement*	*As per Bujharat out of 'As per Settlement'*
(1)	(2)	(3)	(4)	(5)
Jainagar	0.00	0.00	1,865.93 (9.55)	1,790.88 (95.98)
Jharia	277.98 (0.49)	44.13 (15.88)	3,507.12 (4.96)	2,721.80 (77.61)
Katras	368.79 (1.60)	168.84 (45.78)	4,233.52 (18.51)	3,449.27 (81.48)
Nawagarh	849.80 (3.57)	249.42 (29.35)	2,382.31 (7.36)	2,292.75 (96.24)
Nagarkiari	110.71 (0.66)	92.42 (83.48)	912.67 (6.34)	800.10 (87.67)
Pandra	81.01 (0.44)	42.03 (51.88)	4,657.33 (4.43)	3,581.33 (76.90)
Tundi	218.31 (0.93)	18.82 (8.62)	2,976.43 (4.02)	2,445.39 (82.16)
District Total	1,906.60 (1.17)	615.66 (32.29)	20,535.31 (6.06)	17,081.52 (83.18)

Note: Figures in brackets in columns 2 and 4 are the percentages to the lands under the control of the zamindars and tenure-holders given in columns 2 and 3 respectively of Table 5.2 Figures in brackets in columns 3 and 5 are the percentages to the lands under personal cultivation 'As per Settlement' given in columns 2 and 4 respectively.

zamindars and tenure-holders at the time of the survey and settlement and *bujharat*, shows that the zamindars not only did not increase land under their personal cultivation after the survey and settlement, but did not even retain with them whatever they had at that time under their personal cultivation. Out of 1,906.60 acres of land under personal cultivation of the zamindars in the district as a whole at the time of the survey and settlement, only 615.66 acres (32.29 per cent) were under their personal cultivation at the time of the *bujharat*. The zamindar of Tundi estate was left with only 8.62 per cent of the land held by him under his personal cultivation. There were 218.31 acres under personal cultivation of the zamindar of this estate at the time of the survey and settlement, while only 18.82 acres out of that land were in his possession at the time of the *bujharat*. Similar was the fate of the Jharia estate. The total land under personal cultivation of the zamindar of this estate at the time of the survey and settlement was 277.98 acres. At the time of the *bujharat* he had only 44.13 acres (15.88 per cent) left with him under his personal cultivation out of that land. The zamindar of the Katras estate had lost, during the period between the survey and settlement and *bujharat*, 54.22 per cent of land held by him under his personal cultivation at the time of the survey and settlement. All

TABLE 5.6

LAND SOLD AND SETTLED, DURING THE PERIOD BETWEEN THE SURVEY AND SETTLEMENT (1918-1925) AND *BUJHARAT* (1956), BY THE ZAMINDARS AND TENURE-HOLDERS OUT OF THE LAND HELD BY THEM UNDER THEIR PERSONAL CULTIVATION AT THE TIME OF THE SURVEY AND SETTLEMENT

(in acres)

Name of the estate	*Zamindars*			*Tenure-holders*		
	Land under personal cultivation 'As per Settlement'	*Sold out of land under personal cultivation*	*Settled out of land under personal cultivation*	*Land under personal cultivation 'As per Settlement'*	*Sold out of land under personal cultivation*	*Settled out of land under personal cultivation*
(1)	(2)	(3)	(4)	(5)	(6)	(7)
Jainagar	0.00	0.00	0.00	1,865.93	0.00	75.05 (4.02)
Jharia	277.98	124.90 (44.93)	77.47 (27.87)	3,507.12	329.68 (9.40)	227.57 (6.49)
Katras	368.79	42.39 (11.49)	157.56 (42.72)	4,233.52	253.24 (5.98)	524.26 (12.38)
Nawagarh	849.80	441.28 (51.93)	159.10 (18.72)	2,382.31	39.97 (1.68)	49.59 (2.08)
Nagarkiari	110.71	0.00	18.29 (16.52)	912.67	0.94 (0.10)	111.63 (12.23)
Pandra	81.01	29.90 (36.91)	0.95 (1.17)	4,657.33	329.56 (7.08)	615.52 (13.22)
Tundi	218.31	46.64 (21.36)	151.57 (69.43)	2,976.43	90.52 (3.04)	394.12 (13.24)
District Total	1,906.60	685.11 (35.93)	564.94 (29.63)	20,535.31	1,043.91 (5.08)	1,997.74 (9.73)

Note: Figures in brackets in columns 3 and 4 are the percentages to the lands under personal cultivation of the zamindars given in column 2 and figures in brackets in columns 6 and 7 are the percentages to the lands under personal cultivation of the tenure-holders given in column 5.

other estates, except Nagarkiari, show a good agreement, with each other in this respect. It will be seen from the case studies of the financial position of two zamindaris given in the next chapter that the zamindars were always heavily encumbered. The methods by which they reduced their debts and met their requirements for ready money were either to sell their proprietary rights in part of their holdings or to settle their lands with tenants as considerable sums of money by way of *salami*, besides the

TABLE 5.7

LAND BROUGHT UNDER PERSONAL CULTIVATION, DURING THE PERIOD BETWEEN THE SURVEY AND SETTLEMENT (1918-1925) AND *BUJHARAT* (1956), BY THE ZAMINDARS AND TENURE-HOLDERS OUT OF UNCULTIVATED LAND IN THEIR POSSESSION AT THE TIME OF THE SURVEY AND SETTLEMENT

(in acres)

Name of the estate	*Zamindars*		*Tenure-holders*	
	Uncultivated land in possession of the zamindars	*Land brought under personal cultivation out of uncultivated land*	*Uncultivated land in possession of the tenure-holders*	*Land brought under personal cultivation out of uncultivated land*
(1)	(2)	(3)	(4)	(5)
Jainagar	0.00	0.00	6,248.69 (31.99)	601.17 (9.62)
Jharia	14,959.20 (26.24)	20.64 (0.14)	18,113.21 (25.62)	1,996.27 (11.02)
Katras	9,370.50 (40.65)	150.12 (1.60)	6,276.98 (27.44)	554.97 (8.84)
Nawagarh	8,161.65 (34.29)	54.88 (0.67)	7,439.65 (22.97)	519.19 (6.98)
Nagarkiari	4,623.48 (27.40)	0.00	2,713.68 (18.86)	49.00 (1.81)
Pandra	5,718.12 (30.72)	2.50 (0.04)	31,050.94 (29.51)	18.39 (0.06)
Tundi	15,331.58 (65.24)	33.47 (0.22)	34,460.91 (46.56)	265.60 (0.77)
District Total	58,164.53 (35.72)	261.61 (0.45)	1,06,304.06 (31.35)	4,004.59 (3.77)

Note: Figures in brackets in columns 2 and 4 are the percentages to the lands under the control of the zamindars and tenure-holders given in columns 2 and 3 respectively of Table 5 2. Figures in brackets in columns 3 and 5 are the percentages to the uncultivated lands in possession of the zamindars and tenure-holders given in columns 2 and 4 respectively.

annual rental, were realised at each such occasion. As the zamindars had the best lands of the villages under their personal cultivation, it was natural for the prospective buyers or lessees to ask for buying or renting in those lands only. Being needy, the zamindars sold away or settled with others whatever was demanded of them. This will be borne out from Table 5.6 which presents data about the land sold and settled, during the period between the survey and settlement and *bujharat*, by the zamindars and tenure-holders out of the land held by them under their

TABLE 5.8

LAND SOLD AND SETTLED, DURING THE PERIOD BETWEEN THE SURVEY AND SETTLEMENT (1918-1925) AND *BUJHARAT* (1956), BY THE ZAMINDARS AND TENURE-HOLDERS OUT OF UNCULTIVATED LAND IN THEIR POSSESSION AT THE TIME OF THE SURVEY AND SETTLEMENT

(in acres)

Name of the estate	*Zamindars*			*Tenure-holders*		
	Uncultivated land in possession of the zamindars	*Sold out of uncultivated land*	*Settled out of uncultivated land*	*Uncultivated land in possession of the tenure-holders*	*Sold out of uncultivated land*	*Settled out of uncultivated land*
(1)	(2)	(3)	(4)	(5)	(6)	(7)
Jainagar	0.00	0.00	0.00	6,248.69	0.00	869.64 (13.92)
Jharia	14,959.20	851.71 (5.69)	1,862.45 (12.45)	18,113.21	706.63 (3.90)	2,172.71 (12.00)
Katras	9,370.50	245.94 (2.62)	553.80 (5.91)	6,276.98	226.80 (3.61)	1,021.93 (16.28)
Nawagarh	8,161.65	233.59 (2.86)	1,975.24 (24.20)	7,439.65	318.84 (4.29)	1,735.82 (23.32)
Nagarkiari	4,623.48	0.00	1,263.31 (27.32)	2,713.68	0.00	666.29 (24.55)
Pandra	5,718.12	103.74 (1.81)	544.07 (9.51)	31,050.94	327.74 (1.06)	4,040.17 (13.01)
Tundi	15,331.58	0.00	1,871.09 (12.20)	34,460.91	34.83 (0.10)	2,646.59 (7.68)
District-Total	58,164.53	1,434.98 (2.47)	8,069.96 (13.87)	1,06,304.06	1,614.84 (1.52)	13,153.15 (12.37)

Note: Figures in brackets in columns 3 and 4 are the percentages to the uncultivated lands in possession of the zamindars given in column 2 and figures in brackets in columns 6 and 7 are the percentages to the uncultivated lands in possession of the tenure-holders given in column 5.

personal cultivation at the time of the survey and settlement. It shows that out of 1,906.60 acres of land under personal cultivation of the zamindars at the time of the survey and settlement, 1,250.05 acres (65.56 per cent) had been sold away and settled with others by them by 1956. In the case of the Tundi estate, the financial position of which has been studied in the next chapter, the zamindar was in such a hopeless and helpless condition that he gave away during that period 90.79 per cent (21.36 per cent by way of sale and 69.43 per cent by way of settlement) of his land held by him under personal cultivation. In the case of the zamindar of the Katras estate, owning coal-bearing lands, about whose financial position also a study has been made in the next chapter, it was the sub-soil rather than the soil which was more important and fetching huge amounts by way of *salami* and royalty to the zamindar and therefore the percentage of land sold and settled (54.21) by him, during that period, out of the land under his personal cultivation was not as high as in the case of the Tundi estate. All the other estates show a good agreement with this trend.

Table 5.7, which gives information about the new land brought under personal cultivation by the zamindars and tenure-holders, during the period between the survey and settlement and *bujharat*, out of uncultivated land in their possession, shows that in spite of large areas of uncultivated land at their disposal, an insignificant amount of land, that is 261.61 acres (0.45 per cent of the total uncultivated land at the disposal of the zamindars), was added to their land under personal cultivation by all the zamindars put together in the district as a whole during that period. Table 5.8, which gives land sold and settled by the zamindars and tenure-holders out of uncultivated land in their possession, during the period between the survey and settlement and *bujharat*, shows that against that less than half a per cent land brought by the zamindars under their personal cultivation in the district as a whole from their uncultivated land, they had given away 16.34 per cent of their uncultivated land to others during the same period—13.87 per cent (8,069.96 acres) was settled with tenants and 2.47 per cent (1,434.98 acres) was sold by all the zamindars put together in the district out of their 58,164.53 acres uncultivated land. The zamindar of the Tundi estate who had settled with tenants 1,871.09 acres (12.20 per cent) out of 15,331.58 acres of uncultivated land in his possession (*vide* Table 5.8), had brought only 33.47 acres (0.22 per cent) under his personal cultivation during the same period (*vide* Table 5.7) from that land. In the Jharia estate, only 20.64 acres (0.14 per cent) out of its 14,959.20 acres uncultivated land were brought under personal cultivation by the zamindar (*vide* Table 5.7), while 2,714.16 acres (18.14 per cent) of its uncultivated land were settled and sold by the zamindar during that period (*vide* Table 5.8). In the Pandra estate the land sold and settled by the zamindars, during that period, out of their 5,718.12 acres uncultivated land was 647.81 acres (11.32 per cent), while the land brought under their personal cultivation out of that uncultivated land was only 2.50 acres (0.04 per cent). As immediately after the conclusion of the survey and settlement operations in 1925 the Katras estate was brought under the management of the Government under the protection of the Encumbered Estates Act, there was comparatively more new land (1.60 per cent of its uncultivated land) brought under personal cultivation of the zamindar in that estate than in any other estate during that period. As the sales and settlements of land produced immediate results, the broken zamindars neglected their future interests and preferred ready money to bringing land under their personal cultivation. From all the new settlements and sales, large sums in the shape of *salami* must have been realised from the tenants and tenure-holders, besides the usual rental.

Raiyati Land

Table 5.9, which gives *raiyati* land settled by the zamindars and tenure-holders in all the estates as per settlement and *bujharat*, shows that *raiyati* land with the tenants had increased in the district as a whole from 2,83,222.28 acres in 1918-1925 to 3,02,779.86 acres in 1956, an increase of 6.91 per cent. On further examination of the same data it will be observed that it had increased under both the zamindars and tenure-holders. Total *raiyati* land under the zamindars was 95,803.70 acres at the time of the survey and settlement and it stood at 1,05,229.81 acres at the time of the *bujharat*, showing an increase of 9.84 per cent. In the Tundi estate, the land settled with the raiyats by the zamindar had increased by as much as 27.70 per cent during that period. Similarly, the tenure-holders had, under *raiyati*, 1,87,418.58 acres at the time of the survey and settlement and 1,97,550.05 acres at the time of the *bujharat* in the district as a whole. This also shows an increase of 5.41 per cent in the *raiyati* land during the interval between the survey and settlement and *bujharat*. This increase in *raiyati* land under the zamindars and tenure-holders was on account of the land settled with the tenants by the zamindars and tenure-holders during the period between the survey and settlement and *bujharat*. Figures of all the estates except that of Jharia show an increase in *raiyati* land during that period. In the Jharia estate also there was an increase in *raiyati* land under the zamindar during that period but it had declined under the tenure-holders by 4.79 per cent from 39,835.05 acres at the time of the survey and settlement to 37,925.54 acres at the time of the *bujharat*. Actually, there was an increase in *raiyati* land under the tenure-holders of this estate also during that period as will be evident from Tables 5.6 and 5.8 which give information about the land settled, during the period between the survey and settlement and *bujharat*, by the zamindars and tenure-holders out of the land under their personal cultivation and their uncultivated land respectively. This overall decrease in *raiyati* land

TABLE 5.9
RAIYATI LAND AS PER SETTLEMENT (1918-1925) AND *BUJHARAT* (1956)

(in acres)

Name of the estate	*Zamindars*		*Tenure-holders*	
	As per Settlement	*As per Bujharat*	*As per Settlement*	*As per Bujharat*
(1)	(2)	(3)	(4)	(5)
Jainagar	0.00	0.00	10,971.20 (56.16)	11,915.89 (+8.61)
Jharia	38,130.54 (66.88)	39,711.70 (+4.15)	39,835.05 (56.35)	37,925.54 (—4.79)
Katras	12,650.80 (54.88)	13,659.91 (+7.98)	10,194.46 (44.57)	12,225.67 (+19.92)
Nawagarh	13,730.92 (57.68)	16,537.20 (+20.44)	15,994.19 (49.39)	18,011.90 (+12.62)
Nagarkiari	11,608.16 (68.79)	12,889.76 (+11.04)	10,228.76 (71.08)	11,007.62 (+7.61)
Pandra	12,213.87 (65.63)	12,892.53 (+5.56)	65,465.18 (62.21)	68,567.15 (+4.74)
Tundi	7,469.41 (31.78)	9,538.71 (+27.70)	34,729.74 (46.93)	37,896.28 (+9.12)
District Total	95,803.70 (58.83)	1,05,229.81 (+9.84)	1,87,418.58 (55.27)	1,97,550.05 (+5.41)

Note: Figures in brackets in columns 3 and 5 are the percentages increase or decrease to the *raiyati* lands 'As per Settlement' given in columns 2 and 4 respectively. Figures in brackets in columns 2 and 4 are the percentages to the lands under the control of the zamindars and tenure-holders given in columns 2 and 3 respectively of Table 5.2.

under the tenure-holders of this estate was due to the fact that out of 39,835.05 acres of *raiyati* land under the tenure-holders of this estate at the time of the survey and settlement, 4,608.46 acres (11.57 per cent) were acquired by the Government, for different public utility purposes, during the period between the survey and settlement and *bujharat*. Out of these 4,608.46 acres, 1,203.45 acres (26.11 per cent) were acquired for the Bihar Institute of Technology, Sindri, 1,028.85 acres (22.33 per cent) were acquired for the Damodar Valley Corporation, 2,005.69 acres (43.52 per cent) were acquired for the Fertilizers Corporation of India Limited, Sindri, and 370.47 acres (8.04 per cent) were acquired for the National Coal Development Corporation. Besides, 419.43 acres (1.05 per cent) of *raiyati* land held by the raiyats under the tenure-holders of this estate at the time of the survey and settlement had subsided and 71.75 acres (0.18 per cent) had been encroached upon during that period. The *raiyati* land under the control of the zamindar of this estate had also been acquired by the Government for the above-mentioned public utility purposes but the land settled with the raiyats by the zamindar during that period was more than the land acquired by the Government from the raiyats during the same period and therefore there was a net increase in *raiyati* land under the zamindar during that period.

The increase in *raiyati* land during the interval between the survey and settlement and *bujharat* under both the zamindars and tenure-holders was not unexpected. With the passing of the Chota Nagpur Tenancy Act, 1908 (Bengal Act 6 of 1908), and the preparation of the record-of-rights in 1918-1925, the position of the tenants had become very strong; they could not be evicted from their holdings by the landlords except in execution of decrees for ejectment on specific grounds (*vide* section 22 of the Chota Nagpur Tenancy Act). Though, in the beginning, holdings of the tenants were sold in full for arrears of rent, with the passing of the Bihar Money-lenders (Regulation of Transactions) Act, 1939, Bihar Act 7 of 1939, only part holdings sufficient to liquidate the decretal amount were put to sale in execution of certificate cases (*vide* section 14) and therefore the zamindars and tenure-holders could not get back much land out of such sales. Even when they purchased on account of such sales they did not bring those lands under personal cultivation but resettled them with other tenants by realising decent sums of money in the shape of *salami*. There was, however, between 1929 and 1937, sale of land on a somewhat bigger scale for arrears of rent on account of the inability of the tenantry to pay such arrears by reason of the unprecedented fall in prices during the said period. There was an agreement between the zamindars and the popular ministry, formed at that time, by the end of 1937 whereby lands sold in execution of rent decrees between those years were to be restored to the original tenants on payment of half the decretal amount in five equal instalments extending over a period of five years.[26] To give legal effect to this agreement, an Act entitled "The Bihar Restoration of Bakasht Lands and Reduction of Arrears of Rent Act, 1938 (Bihar Act 9 of 1938)", was passed, in 1938, incorporating all the provisions of the above-mentioned agreement. It provided that all raiyats whose holdings or portions of whose holdings were sold at any time between January 1, 1929, and December 31, 1937, in execution of decrees for arrears of rent and were purchased by the landlords of such holdings and were in the possession or under the control of the said landlords, were to be restored to the former tenants in the manner prescribed

[26] *The Indian Nation*, Patna, dated December 22, 1937; and Letter No. 870,II T-127 R., dated February 7, 1938, from the Secretary to the Government, Revenue Department, to the Secretary, Board of Revenue, Bihar (*vide* File No. IIT-127 R. of 1938 of the Board of Revenue, Bihar).

in that Act (*vide* section 3). Section 16 of the same Act also provided that where any arrears of rent were due to a landlord for any period ended before *Asin Badi* 1 of *Sambat* 1994 (1937 A.D.), the raiyat could make an application to the Collector for the reduction of such arrears. Another point favourable to the tenants in Chota Nagpur was that the occupancy-right according to the Chota Nagpur Tenancy Act did not depend on prescriptive right like twelve-year period of occupation mentioned in the Bengal Tenancy Act, 1885, Central Act 8 of 1885 (*vide* section 20) which was in force in the rest of the Bihar State. The Chota Nagpur Tenancy Act saved the customary right of acquisition of occupancy-rights by the tenants immediately on their occupation of land (*vide* section 76). Yet another important feature of the Chota Nagpur Tenancy Act in which it differed from the Bengal Tenancy Act, and was beneficial to the tenants, was the restrictions imposed by it on the transfer of their rights by the occupancy-raiyats in their holdings or portions thereof by (a) mortgage or lease, for any period, expressed or implied, which exceeded or could in any possible event exceed five years, or (b) sale, gift or any other contract or agreement provided that a raiyat could enter into a *bhugatbandha* (usufructuary) mortgage of his holding or any portion thereof for any period not exceeding seven years [*vide* clause (1) of section 46]. In 1947, the above-mentioned section 46 was replaced by a new section [*vide* section 14 of the Chota Nagpur Tenancy (Amendment) Act, 1947 (Bihar Act 25 of 1947)]. The new section 46 allowed the occupancy-raiyats who were not aboriginals or members of the scheduled-castes to transfer, without the sanction of the Deputy Commissioner, their rights in their holdings or portions thereof to other persons who were residents within the local limits of the districts in which the holdings were situated, but prohibited the aboriginal and scheduled-caste raiyats from transferring their rights in their holdings to raiyats who were not aboriginals or were not members of the scheduled-castes. The second proviso to sub-section (1) of the new section 46 provided that an occupancy-raiyat who was an aboriginal or a member of the scheduled-caste could transfer, with the previous sanction of the Deputy Commissioner, his right in his holding or portion of his holding by sale, exchange, gift or will only to another aboriginal (who was resident within the local limits of the area of the police-station within which the holding was situated) or person who was a member of the scheduled caste (residing within the local limits of the district within which the holding was situated) respectively.

Though for the reasons mentioned above the *raiyati* land remained intact in possession of the tenants in the district, the provisions of law restricting the transfer of land by the raiyats, particularly the aboriginal and scheduled-caste raiyats, to other tenants were evaded by allowing the land to be surrendered to the landlords and resettled by them to outsiders on payment of consideration money to the original tenants. As mentioned above, by virtue of the provisions of section 46 of the Chota Nagpur Tenancy Act, a raiyat had got only a limited right of transfer of land from his holding. In order to avoid the restrictions imposed under the law, a circuitous method of transfer was often resorted to by many raiyats in the district. What was done is this. When a purchaser was available to whom the raiyat could not transfer his land in view of the provisions of section 46 of the Chota Nagpur Tenancy Act, the landlord of the holding was approached for accepting the surrender of the land, which was permissible under section 72 of that Act, and on getting a promise from the landlord that the surrendered land would be settled with the intending purchaser the land was surrendered by the raiyat and the intending purchaser then paid the price agreed upon and the landlord on receipt of some consideration by way of *salami* from the vendee settled the land with him. It was clearly an ingenious method of transfer of land, which was not transferable, to defeat the provisions of the law.

The earliest reported decision of the Patna High Court on the legality of such surrenders and resettlements is in the case of *Bario Santhal and others* v. *Fakir Santhal* [*vide* A.I.R. 1924 Patna 793 or 75 (1923) I.C. 601*] in the Appeal No. 1068 of 1921, decided on April 17, 1923, from the Appellate Decree of the District Judge of Manbhum, dated April 14, 1921, by which he had reversed the decision of the *Munsif* (civil judge of the lowest rank) of Chaibasa, Singhbhum district, dated November 13, 1920. In that case certain occupancy-tenants had surrendered their holdings to their landlord on account of partially a very small amount of rent which was due to the landlord from them, but largely in lieu of repayment of money which had been lent by the landlord to the tenants. The *Munsif* had come to the conclusion that the landlord, having admitted that the tenants had surrendered the lands in suit mainly to liquidate the debts due to him, could not defend the transaction in its main elements because it had taken place in contravention of the provisions of section 46 of the Chota Nagpur Tenancy Act. As the consideration for the surrender was liquidation of debts, the *Munsif* held the view that the relinquishment was tendered to the landlord not as *pradhan* (landlord) but as money-lender and the entire transaction was really a sale covered by a camouflage or surrender. As the sale of *raiyati* land was strictly prohibited, he declared the *istifanama* (surrender) by the tenants as invalid in law. But the District Judge was of a contrary opinion and held that the transaction was a surrender in favour of the landlord and not a sale as decided by the *Munsif*. According to him there did not exist a distinction in the law between surrender to the *pradhan* qua *pradhan* and qua money-lender and therefore the view which the *Munsif* had taken was not in law justified. The decision of the District Judge was

*This citation is to be read for the '*All India Reporter*, 1924, Patna Section, page 793, or the *Indian Cases*, Volume 75, 1923, page 601'.

upheld by the Patna High Court which laid down that though it was possible that such surrenders might conceivably result in some form of evasion being effected against the intention of the Chota Nagpur Tenancy Act yet the decisions, however, seemed to be clear that it was open to a tenant to give back his holding, that is surrender it, to his landlord even for a pecuniary consideration, and such a transaction did not amount to a transfer in contravention of the provisions of section 46 of the Chota Nagpur Tenancy Act. In the concluding paragraph of his judgement the judge mentioned about his earlier judgement (unreported) in the case of *Nagru Kumhar* v. *Chhuku Nath Mandal* in the second Appeal No. 1930 of 1921, where he had gone a step further and held that even where a third party had paid consideration to a tenant as a result of which the tenant had agreed with the landlord to surrender his holding whilst the landlord had agreed to resettle the property with the person who had given the consideration to the outgoing tenant, the circuitous arrangement was not illegal since surrender was not a transfer within the meaning of section 46. This view of the Patna High Court was based upon the following three decisions which had been given earlier in that and in other courts. In the Appeal No. 19 of 1916, decided on July 16, 1917, from the Appellate Order of the Sub-Judge of Sambalpur, dated September 21, 1916, in the case of *Trilochan Panda* v. *Dinabandhu Panda*, it had been held by the Patna High Court that a surrender of tenancy in favour of the landlord was not a transfer within the meaning of section 46 of the Central Provinces Tenancy Act, 1898, Central Act 11 of 1898 [*vide* A.I.R. 1917 Patna 103 or 44 (1917) I.C. 317]. In that Act the language of the section was very much the same as that utilised in the Chota Nagpur Tenancy Act, 1908. The second case upon which the Patna High Court relied had been decided by the Calcutta High Court on July 11, 1917, in the case of *Abdul Majid* v. *Hari Charan Haldar and others*, in the Appeal No. 745 of 1917, from the Appellate Decree of the District Judge, 24 Parganas, dated January 10, 1917, in which it had been held that the surrender by a tenant, though for consideration, was operative, and that the provisions of the Transfer of Property Act did not require a registered instrument to evidence the surrender [*vide* 53 (1919) I.C. 17]. The last and the most important case (considered by the Court to be more important than the other two), however, relied on by the Patna High Court in that case, was that of *Jairam and others* v. *Gopikisan and others*, decided by the Nagpur Judicial Commissioner's Court on January 31, 1918, in the case of Second Civil Appeal No. 288 of 1917, from the Decree of the Divisional Judge, Nagpur, dated March 30, 1917, in the Civil Appeal No. 59 of 1916. It had been held there that a surrender of an occupancy-holding for a consideration was not a transfer in contravention of the provisions of section 46 of the Central Provinces Tenancy Act, 1898, Central Act 11 of 1898 [*vide* A.I.R. 1918 Nagpur 178 or 47 (1918) I.C. 32]. In another case also decided by the Patna High Court on April 11, 1933, in the Appeal

No. 1122 of 1930, from the Appellate Decree of the Special Sub-Judge, Chaibasa, Singhbhum district, dated May 2, 1930, in the case of *Jagatbandhu Dutta* v. *Shyam Chand Bhole and others*, the Court did not consider such surrender and settlement illegal [*vide* A.I.R. 1933 Patna 446 or 144 (1933) I.C. 946].

It is astonishing that the attention of the Patna High Court in these cases was not drawn to a Privy Council decision dated December 16, 1916, in the Privy Council Appeal No. 45 of 1914 from the Decree dated May 24, 1911, of the Allahabad High Court, in the case of *Moti Chand and others* v. *Ikram-Ullah Khan and others.* Under the provisions of clause (1) of section 10 of the Agra Tenancy Act, 1901 (United Provinces Act 2 of 1901), every proprietor, whose proprietary rights in a *mauza* (village), or any portion thereof, whether in any share therein or in any specific area thereof, were transferred either by sale in execution of a decree or order of a civil or revenue court, or by voluntary alienation, otherwise than by gift or by exchange between co-sharers in the village, became a tenant with a right of occupancy in his *sir* land (land under personal cultivation of the proprietor himself) and in the land which he had cultivated continuously for twelve years at the date of the transfer. Clause (4) of the above-mentioned section provided that every such tenant shall be called an ex-proprietary tenant and shall have all the rights and be subject to all the liabilities conferred and imposed upon occupancy-tenants by that Act. In clause (2) of section 20 of that Act was enacted that the interest of an ex-proprietary tenant, an occupancy tenant or a non-occupancy tenant was heritable but was not transferable in execution of a decree of a civil or revenue court, or otherwise than by voluntary transfer between persons in favour of whom, as co-sharers in the tenancy, such right had originally arisen, or who had become by succession co-sharers therein. The above-mentioned case of *Moti Chand and others* v. *Ikram-Ullah Khan and others* was in respect of a contract to surrender by the ex-proprietary tenants their occupancy-rights in their *sir* lands the relinquishment of which was forbidden under section 20(2) of the Agra Tenancy Act, 1901. In the Privy Council judgement, it was held that the policy of the Agra Tenancy Act, 1901, was to secure and preserve to a proprietor, whose proprietary rights in a village or in any portion of it were transferred by sale, a right of occupancy in his *sir* lands and in lands continuously cultivated by him for twelve years at the date of the transfer, irrespective of whether he wished it to be secured and preserved to him or not and notwithstanding any agreement to the contrary between him and the transferee. It was further held that the policy of the Act was not to be defeated by any ingenious devices, arrangements or agreements between a vendor and a vendee for the relinquishment by the vendor of his *sir* lands or lands which he had cultivated continuously for twelve years at the date of the transfer. All such devices, arrangements and agreements, according to that judgement, were in contravention of the policy of the Act and were contrary to law and were illegal and void, and could not be enforced by the vendee in any civil court or in any court of revenue [*vide* A.I.R. 1916 Privy Council 59 or I.L.R. (39) 1917 Allahabad 173]. As the provisions of section 20 of the Agra Tenancy Act, 1901, regarding the relinquishment of land were somewhat similar to the provisions contained in section 46 of the Chota Nagpur Tenancy Act, 1908, it appears extremely probable that if this ruling of the Privy Council in a somewhat similar case had come to the notice of the Patna High Court, its decision would have been different.

On the strength of the aforesaid *per incuriam* decision of the Patna High Court, however, such surrenders and settlements were being freely resorted to by many raiyats and landlords in the district.[27] The procedure adopted required co-operation of three parties, namely, the transferor, the transferee and the landlord. The landlord had to agree that he would first accept the surrender and then settle the land with the transferee. The landlord was not obliged to collaborate. If he did not approve the game, he need not have played it. But it is inconceivable that the landlord did not treat such occasions as God-sent opportunities to be welcomed with folded hands as they enabled him to extract his *salami* once again. On the other hand it is equally astonishing that the State Government did not amend the Chota Nagpur Tenancy Act, in the light of the above-mentioned decision of the Patna High Court, to stop the transfer of land by the aboriginal and scheduled-caste raiyats to raiyats other than aboriginals or not belonging to the scheduled-castes in this circuitous method.

After thirty-five years of its prevalence, this *per incuriam* decision was overruled in 1958 by the Patna High Court in the case of *Golap Gadi Goala* v. *Rampariksha Rewani and others* [A.I.R (45) 1958 Patna 553]. In this case an occupancy-tenant had agreed by a registered instrument that in the event of default by him to repay his loan by a particular date, he would surrender 0.79 acres of land in suit to the landlord and further cause it to be settled with his money-lender. When the loan was not repaid by the appointed date, the occupancy-tenant pursuant to the aforesaid agreement surrendered the land to the landlord by a registered deed of surrender dated March 28, 1945, and on the same day the landlord settled the same with the money-lender mentioned in the aforesaid agreement. This arrangement was effected with the prior consent and agreement of the landlord. The premium fixed for this settlement was Rs. 400 out of which Rs. 300 were set off against the dues of the money-lender and Rs. 100 were paid to the landlord as *salami*. The question before the court was whether this arrangement was legal and could be given effect to. It was argued before

[27]Letters No. 466 W., dated April 9, 1945, and No. 526 W., dated April 21, 1945, from the Commissioner, Chota Nagpur Division, to the Secretary, Board of Revenue, Bihar (*vide* File No. 20-849 of 1945 of the Board of Revenue, Bihar).

the court that the surrender and settlement were two independent acts. It was contended that under section 72 of the Chota Nagpur Tenancy Act, 1908, it was lawful for a raiyat to surrender his holding to the landlord and after surrender the landlord was fully competent under section 72(4) of the same Act to settle the land with whomsoever he liked. The court came to the conclusion that the acts of surrender and resettlement considered separately might have the appearance of legality, but in this case they were not independent acts. The entire arrangement consisting of series of acts constituted one whole agreement to which the landlord also was a party. In fact, it was with his prior consent that the entire scheme was designed and agreement reached, the main object of which was to by-pass the statutory provision contained in section 46 of the Act. As there was no direct ruling on this point, the case of *Moti Chand* v. *Khwaja Ikram-Ullah Khan* referred to above, which, though not strictly in point, threw considerable light on this question, was relied upon. It was held that the policy of the Legislature in enacting section 46 of the Chota Nagpur Tenancy Act was, more or less, the same as in the case of section 20 of the Agra Tenancy Act, namely, to secure to the tenants inhabiting the area to which the Chota Nagpur Tenancy Act applied their rights in their occupancy-holdings and to protect them from the avaricious money-lenders. Unsophisticated as the people of that area are, but for the legislation, they would be wiped out by people with superior intellect and bigger purse. Here also, the main object of the arrangement was to effect a transfer of the disputed land to the money-lender in satisfaction of his debt, and since this could not have been done directly because of the prohibition contained in section 46 of the Act, they took recourse to this circuitous arrangement. In the opinion of the judge such a transaction amounted to a clear circumvention of section 46 of the Chota Nagpur Tenancy Act and to lend countenance to such a practice would really have meant judicial recognition of fraudulent contravention of section 46. He further laid down that it was the duty of the court to unravel the mask and to expose the illegality of such transactions. If it were not so, the very object of enacting section 46 would be easily frustrated and the legislative policy behind it thwarted to the great detriment of the people of that area. To support his argument he cited a ruling of the Privy Council in the case of *Surajmull Nagoremull* v. *Triton Insurance Company Limited* (A.I.R. 1925 Privy Council 83, at page 84) in which their lordships had observed that no court could enforce as valid, that which competent enactments had declared shall not be valid, nor was obedience to such an enactment a thing from which a court could be dispensed by the consent of the parties, or by a failure to plead or to argue the point at the outset. A prohibitory enactment was not confined to affording a party a protection of which he might avail himself or not as he pleased.

The result of the *per incuriam* judgement of 1923 of the Patna High Court was that though *raiyati* land remained in the hands of the raiyats, all of it did not remain in the hands of the original ones. How much land changed hands in the circuitous way described above between the passing of the Chota Nagpur Tenancy Act in 1908 and the deliverance of the *per incuriam* judgement in 1923 and during the period of prevalence of that judgement from 1923 to 1958 is under investigation.

It may not be out of place to mention here that one does not come across *per incuriam* judgements delivered by different courts (including the Supreme Court) infrequently. In 1934, in the Miscellaneous Appeal No. 301 of 1933, Civil Revision No. 595 of 1933 and Miscellaneous Appeal No. 37 of 1934, decided on September 12, 1934, from the Appellate Order of the District Judge, Shahabad, dated September 16, 1933, in the case of *Mt. Kabutri and another* v. *Ram Prashad Sahu and others*, it was held by the Patna High Court that "it is the business of the Court to know the law and it is only incumbent upon the litigant to come before the Court and state the facts in respect of which he seeks the remedy. He must ask for such remedies as he requires and there is no obligation on him to inform the Court under which Statute or Rule he is entitled to claim those remedies. If he does not state the Statute or the Rule or if he states a wrong Statute or wrong Rule that in no way relieves the Court of the obligation of itself applying the appropriate Rule or section. A proper recitation of the appropriate section or Rule is not necessary for the obtaining of a remedy" (*vide* A.I.R. 1934 Patna 683). It seems that this view had been expressed by the Chief Justice in some earlier cases also because he mentioned in the above-mentioned judgement that "I have had occasion more than once to protest against the curious superstition that a proper recitation of the appropriate section or Rule is necessary for obtaining a remedy" (*vide* p. 685). Though it has not been possible to find out any authority in support of or against the principle expounded in this judgement yet it is admitted on all hands, and is evident from so many judgements, that the courts expect the parties to the dispute to bring to their notice the law on the subject and do not consider it their duty to find out themselves the correct position of the law (statutory as well as case-law) with respect to the lawsuits before them. Strangely, even the maxim *jura novit curia* does not find place in the legal lexicons prepared in the country.

Performance of the Tenure-holders

Though among the tenure-holders also in each estate there might have been some who burnt their candles at both ends, by and large they protected their properties well as will be seen from the same set of Tables, that is Tables 5.5, 5.6, 5.7, 5.8 and 5.9. So far as the question of performance of the zamindars and tenure-holders upto the time of the survey and settlement is concerned, Tables 5.5, 5.7 and 5.9 show that of the total land under the control of the zamindars and tenure-holders at the time of the survey and settlement, the tenure-holders (a) had more land under their personal cultivation than what the zamindars had (*vide* columns 4 and 2 respectively of Table 5.5); (b) had less uncultivated land than that of the zamindars (*vide* columns 4 and 2 respectively of Table 5.7); and (c) had settled less land with the *raiyats* than that settled by the zamindars (*vide* columns 4 and 2 respectively of Table 5.9). It will be seen from Table 5.5 that out of the land under the control of the tenure-holders at the time of the survey and settlement, 6.06 per cent was under their personal cultivation, while out of the land under the control of the zamindars, only 1.17 per cent was under their personal cultivation in the district as a whole. The tenure-holders of the Katras estate had under their personal cultivation as much as 18.51 per cent of the land under their control at the time of the survey and settlement, while the corresponding percentage for the zamindar of that estate was only 1.60. From the number and types of tenure-holders described earlier, it appears that many of them were small and absentee tenure-holders, as the holders of *brahmotar, debotar, shibotar, lakhiraj, patni, dar-patni, mahatran* and *ijara* tenures were generally absentee rent-receivers. It means that cultivating tenure-holders must have been cultivating even more land in all the estates than what the percentages of different estates indicate.

As regards the performance of the zamindars and tenure-holders during the period between the survey and settlement and the abolition of the zamindari system, it will be seen from Table 5·5 that the tenure-holders of the district had in their possession, at the time of the *bujharat*, 83.18 per cent of the land held by them under their personal cultivation at the time of the survey and settlement. Thus they had lost during this period only 16.82 per cent of their land held by them under their personal cultivation at the time of the survey and settlement. The loss of all the zamindars put together in the district during the same period was 67.71 per cent. The zamindar of the Tundi estate had lost during that period as much as 91.38 per cent of his land held by him under his personal cultivation at the time of the survey and settlement, while the tenure-holders of that estate had lost only 17.84 per cent of their land held by them under their personal cultivation. In the case of the Nawagarh estate, the tenure-holders of that estate were found at the time of the *bujharat* to have retained with them 96.24 per cent of their land held by them under their personal cultivation at the time of the survey and settlement, while the zamindars of that estate were able to hold on to only 29.35 per cent of their land. The figures of all the other estates show a very good agreement with the district figures in this respect. Secondly, it will be seen from Table 5.6 that out of the land held by the zamindars and tenure-holders under their personal cultivation at the time of the survey and settlement, they had sold 35.93 and 5.08 per cent respectively during the period between the survey and settlement and *bujharat*. It means that the tenure-holders had given away only 5.08 per cent of their land, held by them under their personal cultivation, to the sub-tenure-holders created by them during that period, while 35.93 per cent of the land held by the zamindars under their personal cultivation had slipped out of their hands into the hands of the tenure-holders created by them during the same period in the district as a whole. The Nawagarh estate was hit most in this respect. The zamindars of that estate had passed on more than half (51.93 per cent) of their land, held by them under their personal cultivation, to the new tenure-holders created by them, while the tenure-holders of that estate had parted with only 1.68 per cent of their land in favour of the new sub-tenure-holders created by them during that period. Figures of other estates show a good agreement with the district figures in this respect also. Thirdly, the same Table reveals that the tenure-holders had settled 9.73 per cent of their land, held by them under their personal cultivation at the time of the survey and settlement, with the tenants against 29.63 per cent settled by the zamindars out of land under their personal cultivation during the interval between the survey and settlement and *bujharat*. As in the case of land sold, among the tenure-holders of all the estates the tenure-holders of the Nawagarh estate settled the least land (2.08 per cent) with the tenants out of land under their personal cultivation. The tenure-holders of the Tundi estate had settled with the tenants 13.24 per cent of the land held by them under their personal cultivation, while the zamindar of that estate had settled with the raiyats as much as 69.43 per cent of the land held by him under his personal cultivation. The figures of other estates also indicate a similar trend. Table 5.7 shows that the tenure-holders brought 3.77 per cent of their uncultivated land under their own cultivation between the conclusion of the survey and settlement and the abolition of the zamindari system, whereas the zamindars had brought only 0.45 per cent of their uncultivated land under their personal cultivation during the same period. The tenure-holders of the Jharia estate had brought as much as 11.02 per cent of their uncultivated land under their personal cultivation, while the zamindar of that estate had brought under his personal cultivation only 0.14 per cent of uncultivated land in his possession during the same period. The tenure-holders of the Katras and Nawagarh estates having coal-bearing lands also did much better than the zamindars of those estates in this respect. As it has been observed, the zamindars of coal-bearing lands did not even retain with them all the land they already had under their personal cultivation. So their not bringing much new land under personal cultivation is understandable. With the starting of coal-mines by the end of the last century in this area, those zamindars started getting huge sums of money from royalty on the coal extracted from their lands. With the expansion of coal-mines, their receipts from the coal-lands also started rising with the result that their income from land under their personal cultivation became insignificant compared to their income from royalty received by them from the coal-mine lessees. Thus they lost interest in land under their personal cultivation. But with the tenure-holders in those estates the case was just the opposite. As more and more mines were opened, more and more land went out of agricultural use and hence the necessity on their part to bring more and more new land under cultivation. In the case of the Tundi estate, as the tenure-holders also did not bring much land, out of their uncultivated land, under personal cultivation, it seems that the area of land available for cultivation was not much because of the fact that uncultivated land there had been included in the reserved forest area. It was owing to this fact that even the percentage of land sold (given away to the newly created sub-tenure-holders) and settled (with the new tenants) by the tenure-holders of that estate, out of their uncultivated land, was much less than the percentage of land sold and settled by the tenure-holders of other estates out of their uncultivated land. This is evident from Table 5.8. It will also be seen from Table 5.8 that the tenure-holders had sold and settled 13.89 per cent of the uncultivated land in their possession in the district as a whole, while the land sold and settled by the zamindars out of the uncultivated land in their possession was 16.34 per cent.

All these lead to the conclusion that the tenure-holders had managed their properties better as compared to the zamindars. It is not surprising. As mentioned earlier, it seems that most of the tenure-holders had small holdings and therefore they did not part with them unless they were in dire need. Moreover, it appears that unlike the zamindars who were not much educated and were lazy to the backbone, the tenure-holders were men of parts and realised the importance of toiling and moiling. Lastly, the tenure-holders did not have the aristocratic disdain of the zamindars for the management of their business. However, the massive data collected about all the tenure-holders of the district have yet to be processed in detail.

Conclusion

As from an early period the zamindars had started making ducks and drakes of their money, they could not keep their heads above water and were over heads and ears in debt. When their estates reached very serious financial crisis, the money lenders also kept them at arm's length as, by the existence of the Chota Nagpur Encumbered Estates Act, the zamindars had lost much in prestige and credit. When raising more loans became ineffectual, it was indispensable for them to part with their proprietary rights, for considerations, over a part of their estates to meet the embarrassing situation. The result was that though the land of the whole district was settled permanently with the zamindars in 1793, two-thirds of it had passed into the hands of the tenure-holders by 1918-1925. In spite of their huge income from rental and *salami* and in the case of zamindars owning coal-bearing lands from royalty also, they did not spend anything on works of improvement except constructing a few *khas bandhs* or excavating a few *khas* tanks in their estates. All their money was spent on luxury and personal aggrandisement. They also did not show any entrepreneurial approach. Though all the lands under coal-mines in this district were under their proprietorship, they leased them out to various coal companies and did not enter into this business themselves. Even in the field of agriculture, in spite of the great demand for agricultural produce generated by the opening of coal-mines, they did nothing either to improve the agriculture in their estates or increase the land under their personal cultivation. As regards the management of their land by the zamindars and tenure-holders, the performance of the tenure-holders was better because they were not unaware of the fact that, unlike the zamindars who had the protection of the Chota Nagpur Encumbered Estates Act, they would have to lie in the bed they made. Contrary to the general belief that the zamindars always increased land under their personal cultivation by resuming *raiyati* lands, it was found that they not only did not increase land under personal cultivation, but sold away or settled with tenants even their existing lands under personal cultivation to get ready money. *Salami* and not the augmentation of land under personal cultivation seemed to be their main concern.

6

D.C. Wadhwa

Zamindars in Debt*

AS mentioned in the last chapter, the zamindars of the Dhanbad district can be classified as coal (having coal-bearing lands) and non-coal (having no coal-bearing lands) zamindars. The zamindars of the Jharia, Katras, Nawagarh and Pandra estates were coal zamindars while those of the Jainagar, Nagarkiari and Tundi estates were unlucky in this respect. But coal or no coal, the famous writing of M.L. Darling about the Punjab peasant that he was born in debt, lived in debt and died in debt was true *verbatim* in the case of those zamindars, with the exception of the zamindar of the Jharia estate. With the increase of prosperity, they developed inordinate love of display and entered on a course of wasteful extravagance. They vied with each other in their ostentations so much that on one occasion a zamindar of this district purchased a new car to attend the marriage ceremony of another zamindar to make sure that he did not 'feel small' by the side of that zamindar. They were never parsimonious. Large sums of money were spent on social occasions. Births and marriages, in particular, in their families were attended with unusual gaiety. For the marriage of their children rich, respectable and old family connections were sought at a very heavy cost. Huge presents were made. Cars, horses, elephants and carriages were added in numbers to fit in with the magnificent palaces built at the cost of lakhs of rupees and furnished most lavishly. The result of all this reckless spending was that they were always heavily encumbered and in the danger of having their estates sold for default. Before 1859, the local authorities in Chota Nagpur used to take up the affairs of a zamindar approaching the state of bankruptcy and stop all actions of the ordinary courts. Revenue-sale law had never been enforced there.[1] After passing of the Civil Procedure Code in 1859 (Central Act 8 of 1859), when the reign of law became stricter, it was no longer possible for the local authorities to interfere and therefore they were armed with legal powers by the provisions of the Chota Nagpur Encumbered Estates Act, 1876 (Central Act 6 of 1876), which was framed on the analogy of the Oudh Talukdars' Relief Act, 1870 (Central Act 24 of 1870), and was primarily meant to save from ruin old and respectable families of zamindars whose debts and transactions had created heavy litigation. The special features of this Act were that on the application of the zamindars or presumptive heirs and in certain circumstances on the application of the Deputy Commissioners, the estates of the zamindars could be placed in charge of the Government and after the publication of the vesting orders, made under the provisions of section 2 of the Act, all proceedings pending in any court in respect of debts contracted by the zamindars became barred and all processes, executions and attachments for or in respect of those debts became null and void, and, during the management by the Government, the movables of the zamindars and their heirs were rendered non-attachable under process of any civil or revenue court and their persons protected from arrest for or in respect of debts and liabilities incurred by them before bringing their estates under Government management (*vide* section 3). The Chota Nagpur Encumbered Estates Act, therefore, was a piece of exceptional legislation. The main reasons set forth to relieve from encumbrances the estates in Chota Nagpur were the political expediency of avoiding the extinction of big zamindars, who were supposed to have a good deal of territorial authority, and the necessity of protecting the aboriginal tenantry, which had a peculiar relationship with the zamindars, by preventing the penetration of lieu purchasers who, it was feared, would not recognise the customary rights of the people and would exploit the estates for their own benefit by sucking the blood of their prey. As the zamindars of the Dhanbad district, by squandering their money, had started falling, from an early period, into a serious state of indebtedness, all the estates of this district, with the exception of Jharia, were managed by the Government under the protection of this Act for different periods at different times.

The history of the working of this system in the district of Dhanbad, however, shows that the existence of the Act had kept open perpetually a door for perpetration of fraud by the zamindars and their staff and also encouraged the zamindars to plunge into reckless dissipation with impunity knowing full well that their estates would ultimately be saved from the results of their own folly and that they would be rescued from ruin by the

[1]Baden-Powell, B.H., *The Land-Systems of British India*, Oxford, 1892, Vol. I, p. 580.

Government under the provisions of the Act. It prevented a sense of responsibility among the wrong-headed scions of the zamindars who did not care to live prudent lives. Though it gave a right to the heirs to protect their estates from the hands of the zamindars who had run upon a course of wasteful extravagance likely to dissipate their property and thus jeopardise the interests of the heirs, cases were not uncommon when the Wards Department had nursed estates and handed them back in sound financial position only to find the new zamindars as extravagant as the old. Thus the impetus for economic management by the zamindars of their zamindaris was taken away by this Act, with the result that the development of the estates with mineral resources in many of them was hampered. This will be apparent from the following pages which give an account, in brief, of the indebtedness of the two estates in the district, one coal and the other non-coal, namely Katras and Tundi respectively, constructed from the material scattered in various files of the Wards Department of the Boards of Revenue of Bengal, Bihar and Orissa, and Bihar states relating to the management of those estates, for the period they were managed by the Government under the provisions of the Chota Nagpur Encumbered Estates Act and the Court of Wards Act.

KATRAS ESTATE

The Katras estate had some of the best coal-lands in the Jharia coal-fields. However, of all the estates under Government management in this district it was most heavily encumbered. It was under the management of the Court of Wards for fourteen years from 1870 to 1884. The estate was then brought under the protection of the Encumbered Estates Act in 1896-1897 for the liquidation of debts incurred by the zamindar. As early as in 1873, it was involved in debt to the extent of Rs. 2,500.[2] By the time of assumption of the charge by the Government in 1896-1897, the debts had amounted to Rs. 1,62,135, and Rs. 57,424 were incurred during the period of Government management. They all were paid in full before the release of the estate from Government management on August 1, 1907. There was a cash balance of Rs. 31,319-6-6 on the day of the release. The annual rental of the estate at the time of assumption of the charge was Rs. 24,511. With the settlement and resettlement of *raiyati* holdings during Government management, it had increased to Rs. 28,990. The total demand of rent and cesses, including arrears, for the period under Government management was Rs. 3,17,161. Out of this, Rs. 2,81,938, that is 98.4 per cent of the then current demand of Rs. 2,86,555, were collected

[2]Letter No. 2431, dated November 28, 1873, from the Deputy Commissioner, Manbhum, to the Commissioner, Chota Nagpur Division (*vide* Unpublished records of the Government of Bengal, Revenue Department).

and Rs. 7,789 remitted. The balance was reported to be recoverable.[3]

On October 22, 1899, when the estate was under Government management, the Nawagarh Coal Company Limited obtained from this estate a mining lease for coal in respect of twenty-one villages for a period of thirty years with an option of renewal for a further period of thirty years. The royalty payable was four annas per ton of coal extracted, with a fixed minimum of Rs. 8,000 a year. This brought an additional annual income of about Rs. 30,000 to the estate.[4] The Nawagarh Coal Company Limited was later represented by the Barakar Coal Company Limited with Messrs Bird and Company as their Managing Agents.

Though after its release in 1907 the coal-mines in the estate were developed and the income of the estate had considerably increased, the debts of the estate had again started swelling and, by 1921, the zamindar was involved in debt which amounted to Rs. 14,00,000, most of the loans having been contracted at fabulous interest of twenty-four to thirty per cent per annum.[5] Being unsuccessful in getting any further loan from elsewhere, the zamindar arranged to get a big loan from the Nawagarh Coal Company who agreed to give him the loan provided their lease was renewed for a longer period. On March 15, 1921, the Nawagarh Coal Company surrendered their old lease and the zamindar granted to them another mining lease for 999 years for coal in respect of the twenty-one villages, which had already been leased out to them in 1899 for a period of thirty years, and obtained a *salami* of Rs. 5,50,342-8-0. As in the earlier lease executed in 1899, the new lease also stipulated a minimum annual royalty of Rs. 8,000 with royalty at the rate of four annas per ton on all coal extracted during the period of the lease. On the same day the zamindar borrowed from the said Company a sum of Rs. 12,19,017-14-6 carrying interest at the rate of seven per cent per annum. The entire loan was

[3]Report on Wards' and Attached Estates in the Lower Provinces for the year 1907-1908, Calcutta, 1908, p. 28.

[4]Final Report and Return of the Katras Encumbered Estate enclosed with Letter No. 958, dated August 14, 1926, from the Deputy Collector Incharge, Wards and Encumbered Estates, Manbhum, to the Deputy Commissioner, Manbhum (*vide* File No. 20-481 of 1926 of the Board of Revenue, Bihar and Orissa).

[5]Final Report and Return of the Katras Encumbered Estate enclosed with Letter No. 516, dated April 3, 1944, from the General Manager, Wards and Encumbered Estates, Manbhum, to the Deputy Commissioner, Manbhum (*vide* File No. 20-619 of 1944 of the Board of Revenue, Bihar); Letter No. 1759, dated December 31, 1925, from the General Manager, Wards and Encumbered Estates, Manbhum, to the Deputy Commissioner, Manbhum, in the case of the Money Suit No. 1121 of 1924 for Rs. 18,700 in the Court of the Subordinate Judge, Dhanbad (*vide* File No. 20-378 of 1925 of the Board of Revenue, Bihar and Orissa); and Letter No. 1043 W/XX(d)-27, dated May 28, 1925, from the Commissioner, Chota Nagpur Division, to the Secretary, Board of Revenue, Bihar and Orissa, in the case of the money Suit No. 103 of 1925 for Rs. 31,940 in the Court of the Subordinate Judge, Dhanbad (*vide* File No. 20-390 of 1925 of the Board of Revenue, Bihar and Orissa).

secured by a mortgage, in favour of the Company, of the property which formed the subject matter of the lease and the mortgage was accompanied by an agreement whereby the lessee Company was to pay the minimum royalty of Rs. 8,000 to the zamindar in cash and set off the balance of the royalty payable to the zamindar against sums payable by the zamindar to the Company firstly as interest and then as principal on account of the loan advanced to the zamindar. It was further stipulated that if the entire amount of the debt was not repaid within twenty years, that is by March 1, 1941, the property would pass to the Company. Thus, in all, the zamindar obtained on that day a sum of Rs. 17,69,360-6-6 (Rs. 5,50,342-8-0 as *salami* and Rs. 12,19,017-14-6 as loan) from the Company, through their Solicitors Messrs Orr Dignam and Company. Out of this amount the sum of earnest money of Rs. 21,001 which was paid on February 21, 1921, and a previous loan of Rs. 4,00,000 amounting to Rs. 4,19,017-14-6, including interest at the rate of eighteen per cent per annum, taken by the zamindar from the said Company at the time of marriage ceremony of his daughter in June, 1920, was deducted and the balance of Rs. 13,29,341-8-0 was paid to him. Out of this amount Rs. 4,00,000 were misappropriated by the then Dewan (manager) of the estate, who was found guilty by the Calcutta High Court and sentenced to two years' rigorous imprisonment and a fine of Rs. 100.[6] In the course of the criminal proceedings against the Dewan, it came to light that Rs. 80,000 (about seven per cent of the loan amount) were paid by the Dewan to a solicitor from Calcutta as a fee for negotiating the loan with the Nawagarh Coal Company. Messrs Orr Dignam and Company, solicitors of Messrs Bird and Company, who had to do for their side everything what the solicitor of the zamindar had done for his, had got only Rs. 2,000 as their fee in the same transaction.[7] In fact Messrs Orr Dignam and Company had done more for their side than what the solicitor of the zamindar had done for his because the mortgage and lease were drafted by them and the investigations were also done by them. It appears that there were no schedules or settled fees for acting as an attorney. But as Messrs Orr Dignam and Company were reported to be a first class firm, it is obvious that Rs. 2,000 must have been fair and legitimate professional charges of an attorney for the services done in the case. In matters of this kind an attorney was entitled to his fee for legal work done and also could, perhaps, customarily, take a negotiating fee. Whatever was the disreputable custom in such cases, no person purporting to be a solicitor could have entered a

[6]Final Report and Return of the Katras Encumbered Estate enclosed with Letter No. 958, dated August 14, 1926, from the Deputy Collector Incharge, Wards and Encumbered Estates, Manbhum, to the Deputy Commissioner, Manbhum (*vide* File No. 20-481 of 1926 of the Board of Revenue, Bihar and Orissa).

[7]Statement of the solicitor of the zamindar during cross-examination in the Criminal Case No. 259 of 1925 in the Court of the Chief Presidency Magistrate, Calcutta (*vide* File No. 20-480 of 1926 of the Board of Revenue, Bihar and Orissa).

preposterous amount of Rs. 80,000 in his books as a fee for negotiating a loan of twelve lakh rupees. The amount is very high and gives rise to suspicion that his conduct was not above-board. He acted for the zamindar in the transaction but the accounts of the estate do not show that any amount was paid to him. The accounts show no item for the payment of anything to the attorney, not even a legitimate fee. The sum of Rs. 80,000 evidently was not for his fee as the solicitor had the insolence to suggest. It was nothing but a fraudulent and collusive transaction. There must have been some underhand practice between the solicitor and the Dewan and probably some other blackguards. It seems evident that the Dewan and the attorney were mixed up in it as well as in the matter of misappropriation of four lakh rupees. It appears to be a scheme of villainy within villainy. Though in the beginning the solicitor maintained that Rs. 80,000 were given to him on account of his professional fee, in his Letter No. 31, dated April 29, 1926, he wrote to the General Manager, Wards and Encumbered Estates, Manbhum, that "it was an out-and-out payment" to him.[8] As a court witness in the criminal case against the Dewan in the Court of the Chief Presidency Magistrate, Calcutta, in the Case No. 259 of 1925, he stated that the proper remuneration for the work done by him would have been four to five thousand rupees. He further stated during cross-examination that in mortgage matters he was entitled to remuneration as an attorney and if any one promised anything else for his labour, he could take it[9] implying that the amount was paid to him as a present, reward or brokerage and the like. But it has been held in a number of cases that a solicitor is not permitted to accept from his client anything in excess of his professional fee. In the case of *Tomson* v. *Judge* decided on June 25, 1855, in the High Court of Chancery, it was held that a solicitor cannot obtain a gift from his client while the relationship of solicitor and client subsists (*vide* 3 Drew. 306*). In an appeal in the House of Lords against the Decree of the Master of Rolls, dated June 30, 1859, in the case of *Tyrrell* v. *The Bank of England and others*, it was held on February 27, 1862, that in equity a solicitor, as a general rule, should not in any way whatever, in respect of any transaction in the relation between him and his client, make any gain to himself, at the expense of his client, beyond the amount of his just and fair professional remuneration (*vide* 10 H.L. Cas. 26**). In *Morgan* v. *Minett*, it was held by the Chancery

[8]*Vide* File No. 20-480 of 1926 of the Board of Revenue, Bihar and Orissa.

[9]Statement of the solicitor of the zamindar during cross-examination in the Criminal Case No. 259 of 1925 in the Court of the Chief Presidency Magistrate, Calcutta (*vide* File No. 20-480 of 1926 of the Board of Revenue, Bihar and Orissa).

*This citation is to be read for the '*Drewry's Reports, Chancery*, 4 Volumes, 1852-1859, Volume III, 1857, page 306'.

**This citation is to be read for the '*Clark's Reports, House of Lords*, 11 Volumes, 1847-1866, Volume X, 1865, page 26'.

Division of the High Court of Justice that in respect of a gift made by a client to his solicitor, there must be a complete absence of fraud, misrepresentation, or even suspicion, and there must also be a cessation of the confidential relation (*vide* 6 Ch. D. 638*). This principle was followed by full bench of the Madras High Court in the Letters Patent Appeal No. 36 of 1917, decided on October 16, 1917, from the (a) Letters Patent Appeal No. 167 of 1916 and (b) Civil Revision Petition No. 346 of 1915 against the Decree of the Principal District *Munsif* of Masulipatam in the Small Cause Suit No. 2680 of 1914, in the case of *Tenjerla Suryanarayana* v. *Prabhala Subbayya*, where their Lordships refused to enforce an agreement by which a litigant bound himself to pay to the clerk of his *vakil* (pleader) a certain amount for giving special attention to a case entrusted by him to his *vakil* which his *vakil* was bound to see in consideration of his fee, as being contrary to public policy (*vide* A.I.R. 1918 Madras 504 or I.L.R (41) 1918 Madras 471). In all such cases the solicitor is bound to restore the property and becomes a constructive trustee for the client. Even before these judgements, in 1742, in the case of *Saunderson* v. *Glass*, it was observed that if an attorney, *pendente lite*, prevailed upon a client to agree to an exorbitant reward, the court would either set it aside entirely, or reduce it to the standard of those fees to which he was properly entitled (*vide* 2 Atk. 296**). So the least that could have been done in this case would have been to make the solicitor disgorge but nothing of this sort appears to have been done.

On December 19, 1923, the zamindar died leaving two minor sons and one minor daughter. The age of the eldest son was four years and three months. According to the custom of lineal primogeniture the said son had succeeded to the estate and was the minor zamindar. The minor's mother filed a petition on December 30, 1923, saying that the estate was heavily involved in debt amounting to several lakhs of rupees and therefore should be taken under the protection of the Government both under the Court of Wards Act and the Encumbered Estates Act.[10] On January 14, 1924, however, she made another petition requesting that the estate should be taken under Government management under the provisions of the Encumbered Estates Act. She submitted that at the time of the death of the zamindar, the estate was involved heavily in debt which amounted to over seventeen lakhs of rupees. She also alleged that valuable

*This citation is to be read for the '*Law Reports, Chancery Division*, 45 Volumes, 1875-1890, Volume VI, 1877, page 638'.

[10]Petition of the mother of the minor zamindar, dated December 30, 1923, enclosed with Letter No. 25 W., dated January 8, 1924, from the Deputy Commissioner, Manbhum, to the Commissioner, Chota Nagpur Division (*vide* File No. 20-32 of 1924 of the Board of Revenue, Bihar and Orissa).

**This citation is to be read for the '*Atkyn's Reports*, *Chancery*, 3 Volumes 1736-1754, Third Edition, Volume II, 1794, p. 295, at page 297'.

properties of the estate had been leased out to various tenants for nominal rents and many permanent settlements had been made with the result that the income of the estate was confined and there was very little chance of future prospect of any increment. She further submitted that there were various injuries done to the estate by the staff which would be brought to light if a thorough enquiry as to the income and expenditure of the estate for the last fifteen years was made. For the reasons stated above she applied for the protection of the estate under the Encumbered Estates Act so that her minor son could get an advantage of the provisions of the said Act.[11]

On a preliminary enquiry made by the Senior Deputy Magistrate, Dhanbad, it was found that the total current demand of the estate at that time was Rs. 1,60,565-7-1, out of which Rs. 26,451-9-9 were due from agricultural tenants, Rs. 1,14,520-7-7 from coal-lands on account of royalty, Rs. 13,523-5-9 from miscellaneous sources such as tolls from *hat* (market), royalty from stone, forest, house rent, etc., and Rs. 6,070 from *khas* land (land under personal cultivation) of the estate. On the side of liabilities, it appeared that a sum of Rs. 17,14,497-9-8½ was due by the estate. Out of the annual income of Rs. 1,60,565-7-1, the royalty due from the Nawagarh Coal Company amounting to Rs. 93,000 was assigned to the said Company for liquidation of loan of Rs. 13,16,769-2-5 due to them by the zamindar. That left a balance of Rs. 67,565-7-1 out of which Rs. 31,821-9-7 were payable by the estate towards the Government land-revenue, cesses, taxes, etc., leaving a balance of Rs. 35,743-13-6 as net profit of the estate. Out of this the Dewan claimed a pension of Rs. 6,672 a year. Out of the net balance of Rs. 29,071-13-6, as much as Rs. 25,000 per annum were to be paid as interest on loans, other than those from the Nawagarh Coal Company, which amounted to Rs. 3,34,600. In this way it was estimated that practically nothing would be left for repayment of the principal and for meeting the ordinary expenditure of the estate. The *rokad* (cash balance) on December 18, 1923, a day prior to the death of the zamindar, was Rs. 5-2-3. It was considered quite impossible at that stage to ascertain the exact liabilities of the estate and therefore the preparation of a proper scheme was postponed till after the estate was taken under management of the Government and the usual notices were served on claimants.[12]

[11]Petition of the mother of the minor zamindar, dated January 14, 1924, enclosed with Letter No. 221, W/I(d)-1, dated February 8, 1924, from the Commissioner, Chota Nagpur Division, to the Secretary, Board of Revenue, Bihar and Orissa (*vide* File No. 20-32 of 1924 of the Board of Revenue, Bihar and Orissa).

[12]Letter No. 25 W., dated January 8, 1924, from the Deputy Commissioner, Manbhum, to the Commissioner, Chota Nagpur Division, enclosing copies of the Reports of the Senior Deputy Magistrate, Dhanbad, dated January 1, 1924, and the Additional Deputy Commissioner, Dhanbad, dated January 2, 1924 (*vide* File No. 20-32 of 1924 of the Board of Revenue, Bihar and Orissa).

As it was recommended by all the officials that the estate should be taken under the provisions of the Court of Wards Act,[13] the Board of Revenue sanctioned the assumption of the charge of the estate under that Act. But at the same time it suggested that there would be more chance of making the estate solvent under the provisions of the Encumbered Estates Act than under the Court of Wards Act seeing that the debts could not be reduced under the latter Act. As the Board considered it extremely doubtful whether the estate could be made solvent within a reasonable period, the Officiating Secretary wrote to the Commissioner that if on further consideration he or the Deputy Commissioner preferred the application of the Encumbered Estates Act, the Board would be prepared to entertain a proposal for the release of the estate from the management under the provisions of the Court of Wards Act, and assuming charge under the other.[14] With reference to the above-mentioned letter of the Board, the Deputy Commissioner, Manbhum, wrote to the Commissioner, Chota Nagpur Division, saying that the estate was in a perfectly solvent condition and should remain under management under the provisions of the Court of Wards Act. According to him the capitalised value of the estate was two crores of rupees and even a debt of seventeen lakh rupees was negligible. He was also of the opinion that the debt of the Nawagarh Coal Company might be ignored as it was given on a usufructuary mortgage of twenty-one villages and the creditors recouped themselves from the income of those villages which were also leased out to them. He was certain that the debt of the Company could be wiped out before the mortgage matured in 1941. As regards the remaining debts, he suggested that they must be paid off as soon as possible by borrowing money at a reasonable rate of interest, say six per cent.[15] Though the estate continued to be managed under the provisions of the Court of Wards Act, the views of the Deputy Commissioner, judged in the light of after-events, were found to be hopelessly wrong.

After assumption of the management it was found that the affairs of the estate were in a chaotic state and the Government dues such as cesses and income-tax and other dues had not been paid and they were all outstanding. The salary of the staff also had not been paid for about two years. About 200 claims were filed against the estate for Rs. 4,57,670-13-7½ out

[13] Letter No. 58 W/I(d)-1, dated January 16, 1924, from the Commissioner, Chota Nagpur Division, to the Secretary, Board of Revenue, Bihar and Orissa (*vide* File No. 20-32 of 1924 of the Board of Revenue, Bihar and Orissa).

[14] Letter No. 20-32/4, dated February 16, 1924, from the Officiating Secretary, Board of Revenue, Bihar and Orissa, to the Commissioner, Chota Nagpur Division (*vide* File No. 20-32 of 1924 of the Board of Revenue, Bihar and Orissa).

[15] Letter No. 173 W., dated March 13, 1924, from the Deputy Commissioner, Manbhum, to the Commissioner, Chota Nagpur Division (*vide* File No. 20-312 of 1938 of the Board of Revenue, Bihar).

of which a sum of Rs. 1,62,007-12-6 was allowed by the Deputy Commissioner under the powers given to him under the provisions of the Act. The creditors having been dissatisfied with the decisions in claim cases instituted civil suits for the realisation of their dues. Almost all the suits were decreed in full against the estate. The decree-holders took out execution proceedings against the estate, attached immovable property and put up the same for sale. The cash balance of the estate which was in the treasury was also attached by the creditors. The Lakurka (H.V. Low and Company), a valuable colliery of the estate, was attached in execution of the Decree No. 499 of 1925 obtained by a creditor for Rs. 24,358 in the Court of the Subordinate Judge of Dhanbad and a sale proclamation was issued fixing February 15, 1926, as the date for sale. The colliery was, however, freed on February 1, 1925, from attachment on payment by the estate a sum of Rs. 16,000 out of this amount.[16] But the colliery was again attached by another creditor in execution of the Decree No. 602 of 1925 for Rs. 10,057 in the Court of the Subordinate Judge, Dhanbad. In order to have one creditor instead of a hundred and one, all clamouring for repayment at the same time, the General Manager, Wards and Encumbered Estates, tried to raise a loan to pay off immediately all the decree-holders and thereby prevent the sale of any part of the estate property but he found out that the amount covered by the decrees and civil suits was much greater than expected and therefore he could not raise that big a loan.[17] In the meantime the estate was further attached by the decree-holders in the Execution Case No. 737 of 1925 for Rs. 7,671 and in the Execution Case No. 17 of 1926 for Rs. 12,771-3-0, both in the Court of the Subordinate Judge, Purulia.[18] A sum of Rs. 3,09,273-7-8 was required for repayment of debts covered by the decrees and suits and the amounts due to the creditors who were pressing for repayment. From the budget estimate sanctioned for the year 1925-1926, it appears that the estate could afford to pay a maximum amount of Rs. 47,423 only excluding the debt of the Nawagarh Coal Company. According to the scheme of the estate also, the estate could not pay more than Rs. 41,290 per annum towards the repayment of debts other than

[16]Letter No. 1424 W., dated December 19, 1925, from the Deputy Commissioner, Manbhum, to the Commissioner, Chota Nagpur Division (*vide* File No. 20-32 of 1924 of the Board of Revenue, Bihar and Orissa); and Final Report and Return of the Katras Encumbered Estate enclosed with Letter No. 958, dated August 14, 1926, from the Deputy Collector Incharge, Wards and Encumbered Estates, Manbhum, to the Deputy Commissioner, Manbhum (*vide* File No. 20-481 of 1926 of the Board of Revenue, Bihar and Orissa).

[17]Letter No. 1156, dated August 9, 1925, from the General Manager, Wards and Encumbered Estates, Manbhum, to the Deputy Commissioner, Manbhum (*vide* File No. 20-32 of 1924 of the Board of Revenue, Bihar and Orissa).

[18]Letter No. 138 W., dated February 2, 1926, from the Deputy Commissioner, Manbhum, to the Commissioner, Chota Nagpur Division (*vide* File No. 20-32 of 1924 of the Board of Revenue, Bihar and Orissa).

those of the Nawagarh Coal Company.[19]

If the loans were not raised and the debts were not repaid, the creditors who had obtained decrees and instituted suits which were pending in courts would have attached the property of the estate for the realisation of dues as had already been done by a few creditors. In order to avoid the danger of any of the property of the estate being sold, owing to the number of debts, the Deputy Commissioner recommended that the estate be released from the management under the Court of Wards Act and brought under the provisions of the Encumbered Estates Act.[20] Besides the Nawagarh Colliery, the Katras estate had many other coal properties from which it derived royalties. The royalty was received from thirty-one villages and the amount received in one year was Rs. 1,14,520.[21] Therefore it was thought to be distinctly detrimental to the interests of the estate if it was allowed to be split among its creditors. The only way left to save it from the pressing demands for repayment from all the creditors at the same time and to prevent the numerous decree-holders proceeding against the estate was to take it under the protection of the Encumbered Estates Act because under that Act all processes, executions and attachments became null and void and no property could be attached when the estate was under Government management under the provisions of that Act. Thus after pointing out the precarious financial condition of the estate, the Deputy Commissioner again very strongly recommended its release from the Court of Wards Act and its protection under the provisions of the Encumbered Estates Act.[22] Consequently the management of the estate under the Court of Wards Act was relinquished and it was retaken under the protection of the Encumbered Estates Act with effect from February 1, 1926.[23] The Nawagarh Coal Company had in the meantime realised a sum of Rs. 5,80,000 as *salami*

[19]Letter No. 1291, dated September 14, 1925, from the General Manager, Wards and Encumbered Estates, Manbhum, to the Deputy Commissioner, Manbhum (*vide* File No. 20-32 of 1924 of the Board of Revenue, Bihar and Orissa).

[20]Letter No. 1373 W., dated 5th/7th December, 1925, from the Deputy Commissioner, Manbhum, to the Commissioner, Chota Nagpur Division (*vide* File No. 20-32 of 1924 of the Board of Revenue, Bihar and Orissa).

[21]The Preliminary Report of the Senior Deputy Magistrate, Dhanbad, enclosed with Letter No. 25 W., dated January 8, 1924, from the Deputy Commissioner, Manbhum, to the Commissioner, Chota Nagpur Division (*vide* File No. 20-32 of 1924 of the Board of Revenue, Bihar and Orissa).

[22]Letter No. 138 W., dated February 2, 1926, from the Deputy Commissioner, Manbhum, to the Commissioner, Chota Nagpur Division (*vide* File No. 20-32 of 1924 of the Board of Revenue, Bihar and Orissa).

[23]The Bihar and Orissa Government Notification dated March 2, 1926, published in *The Bihar and Orissa Gazette* dated March 10, 1926, Part I, p. 106, enclosed with Letter No. 633 W/XXII(d), dated March 31, 1926, from the Personal Assistant to the Commissioner, Chota Nagpur Division, to the Secretary, Board of Revenue, Bihar and Orissa (*vide* File No. 20-32 of 1924 of the Board of Revenue, Bihar and Orissa).

from six sub-lessees and thus recouped the amount which they had paid to the zamindar as *salami*.[24]

As regards the financial position of the estate during the management under the provisions of the Court of Wards Act, the annual rent and cesses demand (excluding the excess royalty) of the estate was Rs. 49,278 when the estate was brought under management in 1924 and Rs. 54,054 when it was released in 1926.[25] The increase was due to the settlement of the land of the estate during the period under management. The total rent and cesses demand of the estate, for the period it was under management under the provisions of the Court of Wards Act (from 1924 to 1926), including interest and arrears, was Rs. 2,51,968, out of which a sum of Rs. 77,719 was collected and Rs. 1,119 were remitted, leaving a balance of Rs. 1,73,130 at the close of the period of the management to be collected. Out of the balance of Rs. 1,73,130, a sum of Rs. 69,827 was reported as bad and irrecoverable, while that of Rs. 1,03,303 was reported to be good and likely to be realised. The Government land-revenue of Rs. 2,376 payable by the estate during the period under the management was paid in full. An amount of Rs. 48,416 was payable by the estate to the Government on account of cesses during the period under Government management out of which Rs. 30,852 were paid, leaving a balance of Rs. 17,564 only. This amount could not be paid by the estate for want of funds. The current assets of the estate were estimated at Rs. 2,28,108 including Rs. 1,50,000 on account of excess royalty payable by the lessee to the estate, and the liabilities amounted to Rs. 61,818, leaving a balance of Rs. 1,66,290 as surplus for repayment of debts, annually. Out of the surplus, Rs. 1,25,000 were adjusted towards the loan of the Nawagarh Coal Company under the terms of the mortgage deed dated March 15, 1921, and the balance of Rs. 41,290 was proposed to be paid to the other creditors of the estate. But a sum of Rs. 19,034 only could actually be paid to the other creditors for want of funds. There was a cash balance of Rs. 3,630-9-9 to the credit of the estate at the time of its release in 1926 from the management under the provisions of the Court of Wards Act.[26] According to the audit report of the Katras estate for the period it was under management under the provisions of the Court of Wards Act, the debts payable by the estate at the end of the period

[24]Letter No. 917 W., dated August 22, 1938, from the Deputy Commissioner, Manbhum, to the Commissioner, Chota Nagpur Division (*vide* File No. 20-312 of 1938 of the Board of Revenue, Bihar).

[25]Note upon the audit of the accounts of the Katras Wards Estate for the years 1924-1925 to 1926-1927 (*vide* Audit Report No. 191 of 1927-1928, p. 2, Board of Revenue, Bihar and Orissa).

[26]Final Report and Return of the Katras Encumbered Estate enclosed with Letter No. 958, dated August 14, 1926, from the Deputy Collector Incharge, Wards and Encumbered Estates, Manbhum, to the Deputy Commissioner, Manbhum (*vide* File No. 20-481 of 1926 of the Board of Revenue, Bihar and Orissa).

amounted to Rs. 11,98,481.[27] This amount included a sum of Rs. 9,26,968 being the loan balance of the Nawagarh Coal Company bearing interest at seven per cent per annum.

According to the Rules an encumbered estate was to be retained under management for a maximum period of fifteen years, but it had been the exception rather than the rule in this district for an estate to be released within fifteen years of its being taken under Government management. As it was considered impossible to wipe off all the debts of the estate within this period, it was recommended to retain it under the management for seventeen years, that is till 1943-1944, when the debts, after adjustment of all the claim cases filed by the creditors of the estate, were expected to be liquidated.[28]

The mining-lease for coal-lands executed by the zamindar in favour of the Nawagarh Coal Company had landed the estate into extremely unfavourable financial circumstances and hampered the management of the estate to a very great extent in respect of repayment of other dues of the estate. The hands of the General Manager, Wards and Encumbered Estates, were, however, tied by section 9 of the Encumbered Estates Act under which he could inquire into the sufficiency of the consideration, for which the lease was granted, only if the lease was dated within three years immediately preceding the publication of an order for taking the estate under the protection of the Encumbered Estates Act. Had there been no such limitation, the wrong done to the estate could have been corrected as soon as the estate was taken under management under the provisions of the Encumbered Estates Act. The initial mistake was made in 1924 in taking the estate under the provisions of the Court of Wards Act instead of under the Encumbered Estates Act on the basis of the erroneous report of the Deputy Commissioner. If the latter Act had been applied in the very beginning, it would have been possible to cancel the mortgage granted to the Company and the estate would have been free of debts within a period of few years. The General Manager, Wards and Encumbered Estates, was unable to do anything under section 9 but he took action under section 8 of the Encumbered Estates Act in respect of interest which was payable to the Nawagarh Coal Company by the estate. The Company claimed compound interest at the rate of seven per cent with six-monthly rests and the principal amount claimed by them

[27]Note upon the audit of the accounts of the Katras Wards Estate for the years 1924-1925 to 1926-1927 (*vide* Audit Report No. 191 of 1927-1928, p. 2, Board of Revenue, Bihar and Orissa).

[28]Letter No. 1280, dated November 3, 1926, from the General Manager, Wards and Encumbered Estates, Manbhum, to the Deputy Commissioner, Manbhum (*vide* Letter No. 886 W., dated August 17, 1928, from the Deputy Commissioner, Manbhum, to the Commissioner, Chota Nagpur Division, in File No. 20-514 of 1928 of the Board of Revenue, Bihar and Orissa).

included dues on account of the former loan of four lakhs of rupees with interest at the rate of eighteen per cent. The Deputy Commissioner, Manbhum, reduced the interest on the loan of four lakhs of rupees to six per cent and interest on the subsequent loan to five per cent. His order was upheld by the Commissioner, Chota Nagpur Division, but reversed by the Member, Board of Revenue, who raised the interest on both the first and the second loans to seven per cent.[29] The rate of seven per cent per annum was to run from the date of execution of the mortgage deed, that is March 15,1921. The result of this order was that the entire amount of excess royalty continued to be appropriated by the Company in part satisfaction of the interest alone. To be more exact even the interest could not be fully met from year to year and the estate was paying rupees twenty thousand annually as income-tax and cess on the amount of surplus royalty which it never received with the result that the scheme which had been prepared was entirely upset and it became clear that the estate could not be saved unless some drastic action was taken against the Company. The General Manager, Wards and Encumbered Estates, submitted a proposal to the effect that under the provisions contained in section 16 of the Encumbered Estates Act the mortgagee Company should be dispossessed and as lessees they should be made to pay the surplus royalty to the estate in its entirety. But according to the legal opinion, it was not possible to do so for various reasons under the terms of the mortgage deed. It is obvious that the Company had taken the very best legal advice obtainable before the mortgage bond was drawn up. The skill with which it was drawn up deserves great admiration. There is little doubt that the draftsman had in view the possibility that the estate would again be brought under the protection of the Encumbered Estates Act and that he aimed at producing a document which would make it impossible or almost impossible for the authorities to reopen the transaction which had been entered into with the zamindar.

On August 2, 1927, the General Manager, Wards and Encumbered Estates, submitted a revised scheme according to which the estate was to be retained under Government management till 1951-1952, to clear off all the debts. The annual rent-roll of the estate was estimated at Rs. 88,740 and the annual recurring disbursements were Rs. 76,740. There was, therefore, estimated a net balance of Rs. 12,000 per annum for repayment of debts amounting to Rs. 2,81,381 including interest. Thus, according to this scheme, the estate was to be set free from all encumbrances by the year 1951-1952, that is within twenty-five years of its being brought under management under the protection of the Encumbered Estates Act. This

[29]Letter No. 917 W., dated August 22, 1938, from the Deputy Commissioner, Manbhum, to the Commissioner, Chota Nagpur Division (*vide* File No. 20-312 of 1938 of the Board of Revenue, Bihar).

scheme was prepared on the assumption that the Nawagarh Coal Company would not be removed from the enjoyment of royalty in respect of the mining-lease for coal-lands of twenty-one villages. It was calculated that the assigned debts due to the Nawagarh Coal Company would be paid off by 1949-1950.[30] But a comparative statement prepared in the office of the Commissioner, Chota Nagpur Division, showed that the total annual receipts came to Rs. 81,338 and the total annual expenditure to Rs. 86,634, leaving an annual deficit of Rs. 5,296 instead of a surplus of Rs. 12,000 as estimated by the General Manager, Wards and Encumbered Estates. Thus it was clear that no feasible scheme could be made out unless a portion of the estate was sold.[31] The mother of the minor zamindar did not have any objection to selling the bulk of the estate in order to clear off its liabilities provided a minimum required for the maintenance of the minors and their dependents 'consistent with their rank and prestige' was sanctioned. But she vehemently opposed the idea of releasing the estate from Government management as such a step at that moment, and before the liabilities had been cleared off, would have, according to her, spelt utter ruin of the estate.[32] In view of the consequences of releasing the estate then vividly set forth in the representation submitted by the mother of the minor zamindar, it was decided to keep the estate under Government management even if it took much longer than twenty-five years to liquidate all the debts.

In the final scheme prepared by the General Manager, Wards and Encumbered Estates, it was mentioned that even in the absence of savings for repayment of debts other than those of the Nawagarh Coal Company, it would be possible to release the estate in twenty-eight years against twenty-five years as provided in the earlier scheme. The principal debts other than those of the Nawagarh Coal Company were given as Rs. 2,32,956. The amount of interest in the next twenty-three years was calculated at Rs. 2,26,573 at the rate of Rs. 9,851 per year. Accumulated interest up to that date was Rs. 38,570. The total amount of interest to be paid in twenty-three years was calculated at Rs. 2,65,123. Thus the total debt to be cleared off was estimated at Rs. 4,98,079 (Rs. 2,32,956 principal and Rs. 2,65,123 interest). It was estimated that this debt could be repaid

[30]Final Scheme of the Katras Encumbered Estate enclosed with Letter No. 858, dated August 2, 1927, from the General Manager, Wards and Encumbered Estates, Manbhum, to the Deputy Commissioner, Manbhum (*vide* File No. 20-514 of 1928 of the Board of Revenue, Bihar and Orissa).

[31]Letter No. 550 W/XIX-9, dated April 12, 1928, from the Officiating Commissioner, Chota Nagpur Division, to the Deputy Commissioner, Manbhum (*vide* File No. 20-514 of 1928 of the Board of Revenue, Bihar and Orissa).

[32]Letter No. 8, dated July 28, 1928, from the mother of the minor zamindar to the Deputy Commissioner, Manbhum (*vide* File No. 20-514 of 1928 of the Board of Revenue, Bihar and Orissa).

in five years from the year 1949-1950, when the debts of the Nawagarh Coal Company would have been cleared off and the estate would have got the royalty from them.[33] This scheme was sanctioned by the Government.[34]

But the estate was released and restored to the possession of its zamindar with effect from January 20, 1941,[35] much in advance of its scheduled date. The circumstance which led to the release of the estate prior to its scheduled date and sanctioned scheme was its inability to meet its expenditure out of its collection on account of an abnormal increase in income-tax and super-tax which rose to Rs. 42,000 against a scheme provision of Rs. 15,000 a year and resulted in a heavier deficit budget of the estate. The Government land-revenue and taxes alone amounted to Rs. 59,566 a year against a total expected income of Rs. 61,169. Thus the estate was unable to meet any other expenditure except Government land-revenue and taxes noted above. So, in order to carry on, there was no other alternative left to the estate but to obtain a loan the raising of which was sanctioned by the Divisional Commissioner.[36] But all efforts to secure a loan failed as nobody agreed to advance any loan to the estate owing to its hopeless financial position. As the condition of the estate was going from bad to worse, there was a proposal to sell the proprietary right of the coal-lands of the villages leased out to the Nawagarh Coal Company. At that time the revenue received by the estate from royalties paid by the Nawagarh Coal Company under the 999-year lease was less than the interest which was due to the Company on the loan taken by the estate from them. The Nawagarh Coal Company were willing to wash out the loan and the mining-lease if they received a *mukarrari* (permanent surface and underground) lease of the entire property, that is twenty-one villages. They were prepared to pay Rs. 20,000 as *salami* plus a dead rent of Rs. 3,000 every year instead of the minimum royalty of Rs. 8,000—land-revenue, cesses, etc., to be deducted from this amount. The Chief Inspector of Mines estimated that in the twenty-one villages there were 246 million tons of

[33]Letter No. 686, dated May 5, 1928, from the General Manager, Wards and Encumbered Estates, Manbhum, to the Deputy Commissioner, Manbhum (*vide* File No. 20-514 of 1928 of the Board of Revenue, Bihar and Orissa).

[34]Letter No. Sl. 2, dated October 19, 1928, from the Secretary, Board of Revenue, Bihar and Orissa, to the Commissioner, Chota Nagpur Division (*vide* File No. 20-514 of 1928 of the Board of Revenue, Bihar and Orissa).

[35]The Bihar Government Notification No. 164 W., dated February 3, 1941, published in *The Bihar Gazette* dated February 12, 1941, Part I, p. 60 (*vide* File No. 20-440 of 1940 of the Board of Revenue, Bihar).

[36]Memo No. 954 W., dated September 6, 1940, of the Deputy Commissioner, Manbhum, referred to in the Final Report and Return of the Katras Encumbered Estate enclosed with Letter No. 516, dated April 3, 1944, from the General Manager, Wards and Encumbered Estates, Manbhum, to the Deputy Commissioner, Manbhum (*vide* File No. 20-619 of 1944 of the Board of Revenue, Bihar).

coal. The royalty payable on that amount of coal at four annas a ton would have been Rs. 61.5 millions. The surface rent of those villages amounted to Rs. 5,372-10-1 a year. The consideration which the Company was prepared to offer to obtain the *mukarrari* over that property was ridiculously small and therefore a transaction on those lines was considered to be inequitable to the estate.[37] On the other hand, the zamindar, who had attained majority by then, objected to that sale before the Board of Revenue and was anxious to administer his estate himself.[38] In view of this and the fact that the estate was hopelessly insolvent the Government decided not to offer to the zamindar a shirt of Nessus and directed that the estate should be released forthwith from management.

The total debt (excluding the amount adjusted towards the loan of the Nawagarh Coal Company) ascertained by determination of claim cases filed against the estate was Rs. 2,74,358 and the amount ascertained and interest accrued on principal debt during the period under Government management (from 1926 to 1941) was Rs. 1,19,206. Thus in all Rs. 3,93,564 only were due by the estate out of which Rs. 84,305 were actually paid to the creditors, Rs. 10,138 were written off during the period under management and Rs. 2,99,121 remained as balance at the time of release of the estate. The amount of royalty adjusted during that period was Rs. 11,87,734-15-9, leaving still a balance of Rs. 11,36,987 to be paid by the estate to the Company. The annual rent and cesses demand (excluding the excess royalty) of the estate was Rs. 54,054 when the estate was taken under management under the protection of the Encumbered Estates Act and Rs. 45,638 when it was released. The decrease was due to the rent and cess reduction operations in the estate during the period under management. The total rent and cesses demand (including interest and arrears) of the estate for the period it was under management under the provisions of the Encumbered Estates Act was Rs. 21,24,550 out of which Rs. 7,32,586 were collected and Rs. 1,59,035 were remitted, leaving a balance of Rs. 12,32,929 at the close of the period of management. Out of the balance of Rs. 12,32,929, a sum of Rs. 10,369 was reported to be bad and irrecoverable and a sum of Rs. 12,22,560 was reported to be good and in the course of realisation. The total annual income of the estate was Rs. 65,017 against an expenditure (including taxes) of Rs. 72,692, that is the expenditure exceeded the income by Rs. 7,675 a year at the time of its release.[39]

[37]Letter No. 917 W., dated August 2, 1938, from the Deputy Commissioner, Manbhum, to the Commissioner, Chota Nagpur Division (*vide* File No. 20-312 of 1938 of the Board of Revenue, Bihar).

[38]Letter No. 13, dated November 22, 1940, from the zamindar to the Secretary, Board of Revenue, Bihar (*vide* File No. 20-440 of 1940 of the Board of Revenue, Bihar).

[39]Final Report and Return of the Katras Encumbered Estate enclosed with Letter No. 516, dated April 3, 1944, from the General Manager, Wards and

To summarize, the debts of the estate in 1873 were Rs. 2,500. In 1896-1897 when the estate was brought under Government management, the debts amounted to Rs. 1,62,135. After coal was discovered and developed in the estate, the zamindar jumped to great wealth in the late nineties of the last century and with that swelled his debts also. At the time of its release from management under the provisions of the Court of Wards Act in 1926, the debts payable by the estate amounted to Rs. 11,98,481. Out of this amount Rs. 9,26,968 were payable to the Nawagarh Coal Company and Rs. 2,71,513 to other creditors. After the management of the estate for fifteen years (from 1926 to 1941) under the protection of the Encumbered Estates Act, the debts at the time of its release in 1941 were reported to be Rs. 11,36,987 due to the Nawagarh Coal Company and Rs. 2,99,121 to other creditors. That was, in short, the hard lot of the Katras encumbered estate which was likely to lose valuable property worth over sixty-one million rupees in lieu of a loan of 1.2 millions only advanced to the zamindar by the Nawagarh Coal Company in 1921. The mortgagor had no doubt the right to redeem the mortgage but under the then existing conditions it was only an illusory dream to entertain a hope of this nature. How the young zamindar managed his zamindari from its release in 1941 till its vesting in the Government at the time of the abolition of the zamindari system in 1954 is currently under study.

TUNDI ESTATE

The Tundi estate was the biggest of the non-coal zamindaris. The Tundi family was closely connected by descent, intermarriage and tradition with those of the Nagarkiari, Nawagarh and Pandra estates.

If the Jharia, Katras, Nawagarh and Pandra estates were rich in coal deposits, the Tundi estate had a great forest wealth, having about 62,000 acres of its total area of 97,508.71 acres (63.6 per cent) under forest. But the successive zamindars of this estate also saddled it with liabilities which it was not able to bear with the result that as early as on March 28, 1887, the entire estate was attached in execution of a decree for a trifling sum of Rs. 6,729 and the decree-holder had applied for the issue of a notice on the zamindar for proclamation of sale.[40] Coming events cast their shadows before them. In December, 1882, that is much before the attachment, the eldest son of the zamindar had approached the Government with the request that the estate should be taken under Government management

Encumbered Estates, Manbhum, to the Deputy Commissioner, Manbhum (*vide* File No. 20-619 of 1944 of the Board of Revenue, Bihar).

[40]Letter No. 66 W/1887-88, dated May 11, 1887, from the Deputy Commissioner, Manbhum, to the Commissioner, Chota Nagpur Division (*vide* Unpublished records of the Government of Bengal, Revenue Department).

under the protection of the Encumbered Estates Act as his father was dissipating the property. On an enquiry made by the local officials it was found that the debts of the estate, both secured and unsecured, amounted to not less than Rs. 60,000 and the interest on the unsecured debts amounted to about Rs. 6,000 a year. The annual surplus available for payment of this interest and the principal was calculated to be Rs. 5,200. As it was felt certain that the zamindar would contract new loans, the local officials recommended the management of the estate under the provisions of the Encumbered Estates Act.[41] But the proposal was disallowed by the Commissioner, Chota Nagpur Division, on the ground that no political grounds had been shown that warranted such an action.[42]

In August, 1884, however, the same son of the zamindar made another application with the same request alleging that his father had created further encumbrances on the estate and was allowing the land-revenue to fall into arrears.[43] This once again set the enquiry in train. When asked to submit a statement of the debts paid off by him during the last three years, the zamindar informed that out of a total debt of Rs. 82,969-4-0 he had paid Rs. 43,382-8-0. This amount had been paid by him not in cash or kind out of the income of the estate but by giving away part of his estate in *mukarrari* leases and usufructuary mortgages, and by giving receipts for advance payment of rents. For the payment of this amount of Rs. 43,382-8-0, he had given forty-one villages, with an aggregate rental of Rs. 3,582 a year, in *mukarrari*, to various persons for a consideration money of Rs. 20,559. He had obtained a sum of Rs. 20,639-8-0 by usufructuary mortgages and Rs. 2,184 were received by him by giving advance rent receipts. From the statement submitted by the zamindar it also appeared that during that period neither any new land had been brought by him under his personal cultivation nor he had settled any new land with the raiyats and therefore there had been no increase in the income of the estate.[44] The result of the enquiry conducted by the officials more or less confirmed the statement of the zamindar. As regards the liabilities of the estate, the result of the detailed enquiry showed that the (a) debts secured on *mukarrari* leases and usufructuary mortgages of which both the principal and interest were to be liquidated from the rents of the assigned lands

[41]Letter No. W/468, dated August 28, 1883, from the Deputy Commissioner, Manbhum, to the Commissioner, Chota Nagpur Division (*vide* Unpublished records of the Government of Bengal, Revenue Department).

[42]Letter No. 66 W/1887-88, dated May 11, 1887, from the Deputy Commissioner, Manbhum, to the Commissioner, Chota Nagpur Division (*vide* Unpublished records of the Government of Bengal, Revenue Department).

[43]*Ibid*.

[44]Letter No. R/398, dated February 1, 1888, from the Sub-Divisional Officer, Gobindpur, to the Deputy Commissioner, Manbhum (*vide* Unpublished records of the Government of Bengal, Revenue Department).

amounted to Rs. 40,197; (b) debts obtained on usufructuary mortgages of which only the interest was to be liquidated from the rents of the assigned lands amounted to Rs. 20,890; (c) unsecured debts (including arrears of interest) bearing interest at rates varying from twelve to twenty-five per cent per annum amounted to Rs. 15,958; and (d) debt due on a decree under execution was Rs. 6,729. Thus the total liabilities of the estate were estimated at Rs. 83,774. From this amount were deducted Rs. 20,000 the amount of *salami* exacted by the zamindar from various persons to whom forty-one villages had been given in *mukarrari*. This amount of *salami* was set off against debts due to the lessees. The outstanding debts, therefore, were calculated at Rs. 63,774. This figure excluded considerable sums which were barred by limitation and other debts of which no satisfactory evidence was produced at that time but was forthcoming afterwards. This enquiry though far more minute and elaborate than that made in 1883 nevertheless went far to confirm the conclusions then arrived at. In 1883, the debts of the zamindar were estimated at Rs. 60,000. Since then he was known to have contracted new debts to the extent of Rs. 13,000 which would bring up the total to Rs. 73,000. The second enquiry placed them at Rs. 83,774 including those debts which had been extinguished by granting *mukarrari* leases. The difference was comparatively small when the treacherous nature of the data is considered. The annual income and expenditure of the estate at the time of the second enquiry were Rs. 19,253 and Rs. 20,423 respectively, leaving a deficit of Rs. 1,170 per annum. The expenditure included an annual payment of Rs. 5,369 and Rs. 2,722 on account of the debts mentioned under (a) and (b) above respectively. The details of the annual receipts and disbursements of the estate revealed that not only nothing was being paid towards the liquidation of either the unsecured loans or the decretal amount, even no provision had been made for the payment of interest on those amounts. It was calculated by the Deputy Commissioner that by the time the secured debts had been cleared off and the assigned rents released, the arrears of interest on the unsecured debts would have amounted to at least Rs. 60,000. Thus even if those assignments in liquidation of the principal and interest were maintained, and if no fresh debts were contracted, the indebtedness of the estate would have increased. Secondly, it had also come to the knowledge of the local officials during the course of the enquiry that the zamindar was negotiating to secure a loan of Rs. 60,000 carrying interest at the rate of ten per cent per annum in return for an *ijara* lease of the entire estate. It was probably not a matter of coincidence that the figure of outstanding debts arrived at after the detailed enquiry agreed almost exactly with the amount of loan which the zamindar was trying to obtain by leasing out his entire estate. The fact of leasing out of his entire estate by the zamindar was considered to be of grave significance. The Government officials were of the view that such a loan would have been tantamount to a permanent transfer of the estate to an absentee landlord working on a purely business footing and with little inducement to respect the traditional privileges of the Santal and other tenants. The prospective lessee was interested only in developing tusser industry in the estate. As a result of that enquiry, it became clear to the officials that the zamindar was continuously adding to his debts rather than taking steps to wipe them out and was granting *mukarrari* leases and usufructuary mortgages to repay his creditors. The leases of that kind were nothing else but alienations, permanent or temporary, of portions of the estate and constituted the very evil which the Encumbered Estates Act was meant to avert. Thus the Deputy Commissioner, Manbhum, strongly recommended the bringing of the estate under the operation of the Encumbered Estates Act. In the scheme prepared by him it was stated that if the estate was brought under Government administration the liabilities would be cleared off within twelve years or a little less as all interest in excess of twelve per cent would, in accordance with the usual practice, be disallowed. According to the scheme, the annual income of the estate was estimated at Rs. 19,253 and expenditure at Rs. 14,248, leaving a surplus of Rs. 5,005 per year to repay the debt of Rs. 63,774 in rather less than twelve years, allowance being made for the progressive reduction in the annual charge for interest.[45] After a protracted correspondence on the subject among different officials, the Lieutenant-Governor accorded his sanction to the proposal on May 17, 1888,[46] and the estate was brought under the protection of the Encumbered Estates Act on June 11, 1888,[47] to avoid the danger of its being broken and sub-divided among a number of petty proprietors.

The debt due by the estate at the time it was brought under Government management amounted to Rs. 80,887 which increased during the period of management by Rs. 51,485 on account of interest, making a total of Rs. 1,32,372. It took twenty years' management to clear off this debt. The estate was made over to the zamindar on August 1, 1907, free from all encumbrances and in a very good condition. The rental of the estate was Rs. 18,476 per annum at the time when it was brought under protection.

[45]Letter No. 66 W/1887-88, dated May 11, 1887, from the Deputy Commissioner, Manbhum, to the Commissioner, Chota Nagpur Division (*vide* Unpublished records of the Government of Bengal, Revenue Department).

[46]Letter No. 1697/766 L.R., dated May 17, 1888, from the Officiating Under Secretary to the Government of Bengal, Revenue Department, to the Secretary, Board of Revenue, Land Revenue Department, Bengal (*vide* Unpublished records of the Government of Bengal, Revenue Department).

[47]Notification dated May 17, 1888, issued by the Commissioner, Chota Nagpur Division, published in *The Calcutta Gazette* dated June 11, 1888, enclosed with Letter No. 1903/868 L.R., dated June 11, 1888, from the Officiating Under Secretary to the Government of Bengal, Revenue Department, to the Secretary, Board of Revenue, Land Revenue Department, Bengal (*vide* Unpublished records of the Government of Bengal, Revenue Department).

It had risen to Rs. 20,940 by the date of its release, owing to the new settlements with the raiyats. During the latter part of the management, fifty-two Santali villages of the estate were also settled. That would have further increased the income of the estate. The total demand of rent and cesses due to the estate for the period it was under Government management (from 1888 to 1907), including Rs. 8,714 arrears due at the time of assumption of the charge, was Rs. 3,80,453 of which Rs. 3,48,825 (including excess collection of Rs. 41) or 93.8 per cent of the then current demand (Rs. 3,71,739) of the period, were collected and Rs. 12,585 remitted, leaving an outstanding balance of Rs. 19,043, of which Rs. 16,808 (88.3 per cent) were reported to be good and in the process of recovery.[48]

But soon after the estate was released in a sound financial condition, it was managed in such a way by the new zamindar as to be heavily encumbered and on the verge of bankruptcy. This came to the notice of the Deputy Commissioner, Manbhum, by a letter dated July 19, 1914, from the Secretary of the Faridpur Loan Office Limited, East Bengal (now Bangladesh), that the zamindar had proposed to raise a loan of a lakh of rupees from that office on the mortgage of his estate.[49] As there was no unusual event in the family or in the management of the estate during that period requiring a large expenditure, the fact that the zamindar, within a short period of seven years, had incurred liabilities for which he required so large a loan, equal to five years' rental, as rental was only Rs. 20,940 at the time of its release, on the security of his estate, afforded sufficient reason to conclude that it could only be on account of reckless extravagance on his part. It would have permanently involved the estate in liabilities which it would not have been in a position to liquidate. It may not be out of place to mention here that the above-mentioned new zamindar had been vigorously pleading in the early eighties of the last century, as a reversionary heir to the late zamindar, for bringing the estate under the protection of the Encumbered Estates Act on the ground that it was rapidly being dissipated by his spend-thrift father.

On the receipt of the above-mentioned letter from the Faridpur Loan Office Limited, the Deputy Commissioner, Manbhum, made some enquiries. It was reported to him by the Sub-Divisional Officer, Dhanbad, that the estate was involved in serious embarrassment. The zamindar of the estate was reported to have borrowed at that time Rs. 3,000 from a pleader in Dhanbad on the security of twenty villages, and it was only after he had arranged that loan that he could pay the road cess arrears for which a warrant had been issued by the Certificate Officer of Purulia after frequent

[48]Report on Wards' and Attached Estates in the Lower Provinces, for the year 1907-1908, Calcutta, 1908, p. 28.

[49]Letter dated July 19, 1914, from the Secretary, Faridpur Loan Office Limited, Faridpur, to the Deputy Commissioner, Manbhum (*vide* File No. 20-271 of 1915 of the Board of Revenue, Bihar and Orissa).

adjournments of the certificate proceedings.[50] The zamindar, when questioned, could not explain to the Deputy Commissioner as to why he had to borrow from a pleader to pay his public dues, or why it had been necessary for him to apply for so large a loan to Eastern Bengal money-lenders. The zamindar also failed to produce a statement of all his debts and liabilities to the Deputy Commissioner. As the Deputy Commissioner was satisfied that the zamindar was heavily encumbered, he recommended that the provisions of the Encumbered Estates Act might be applied to the estate.[51] On further enquiries, it was learnt that the debts of the zamindar amounted to Rs. 1,50,000. The General Manager, Wards and Encumbered Estates, Manbhum, had estimated the interest on those debts at Rs. 23,000 per annum.[52]

In the last week of October 1914 barely a month-and-a-half after the Deputy Commissioner had recommended the bringing of the estate under Government management, a search in the Registration Office showed that the zamindar had recently alienated twenty villages on *chirasthai jot pattas* (permanent cultivating leases). In sixteen of those villages, the annual rental reserved in the leases was the same as the then current rents though the *salami* paid amounted to Rs. 34,759-4-3, a sum which, it is quite clear, was not possible to recover by lawful means on the part of the lessees as the area of waste-land available in those villages was not sufficient to provide that. In one village the annual rental reserved was Rs. 236-4-0, slightly more than the then current demand, and in addition there was a *salami* of Rs. 3,071. The total current annual rental of the remaining three villages at that time was Rs. 350 but the rental reserved in the leases was Rs. 70 only, leaving a profit of Rs. 280, but a *salami* of Rs. 34,997 was shown as having been paid for them.[53]

Those permanent cultivating leases gave the lessees the right to rear lac and tusser cocoons and coke fuel or wood for building homes but made no mention of any such rights being left for the tenants, and its omission was considered sure, by the Commissioner, Chota Nagpur Division, to lead to the tenants being deprived of their rights by the lessees. Practically every

[50]Letter dated August 17, 1914, from the Sub-Divisional Officer, Dhanbad, to the Deputy Commissioner, Manbhum (*vide* File No. 20-271 of 1915 of the Board of Revenue, Bihar and Orissa).

[51]Letter No. 1108 W., dated September 8, 1914, from the Deputy Commissioner, Manbhum, to the Commissioner, Chota Nagpur Division (*vide* File No. 20-271 of 1915 of the Board of Revenue, Bihar and Orissa).

[52]Letter No. 1365 W., dated November 25, 1914, from the Deputy Commissioner, Manbhum, to the Commissioner, Chota Nagpur Division (*vide* File No. 20-271 of 1915 of the Board of Revenue, Bihar and Orissa).

[53]Letter No. 667 W/IX-1, dated 5th/11th May, 1915, from the Commissioner, Chota Nagpur Division, to the Secretary, Board of Revenue, Bihar and Orissa (*vide* File No. 20-271 of 1915 of the Board of Revenue, Bihar and Orissa).

tree capable of producing lac and tusser was already in the possession of some tenant and a tenant paid a tax of six annas per man irrespective of the number of trees he reared lac or tusser on. To lease those trees to the *jot patta* lessees meant that those tenants who were already in possession of those trees would be dispossessed. The lessees of the villages were also given exclusive rights to the cultivation of waste-lands. As a result of that the tenants were going to be deprived of all the grazing lands. In the case of some of the villages leased, the post of *ijardar* had been held by the ancestors of the then *ijardars* for generations, and the introduction of outsiders as *ijardars*, even if the zamindar had the legal power to do so, was going to be greatly resented by the then existing *ijardars* and by the villagers generally, and was considered fairly certain to lead to agrarian trouble. A large number of tenants were Santals who were not likely to tacitly put up with what they considered to be an infringement of their rights.

Apart from the above-mentioned objections to the leases, they appeared to be collusive. They were all granted after the Deputy Commissioner had submitted his recommendation that the estate should be brought under the protection of the Encumbered Estates Act. They were not registered at Dhanbad, where they would be expected to be registered, but at Purulia. Though the leases were granted to seven different persons, it is reported that they all belonged to two families. Thus the aforesaid *chirasthai jot pattas* for all the twenty villages were objectionable in every respect. It is obvious that they were granted to the creditors of the estate, or to the members of their families, with the object of showing that the zamindar had succeeded in raising money to clear off his debts and that consequently there was no longer any reason to take the estate under Government management under the provisions of the Encumbered Estates Act.

It was also reported that the zamindar was making arrangements to lease out to the Bengal Coal Company the whole of timber, bamboo and brush-wood in his estate, with the exception of one or two species of trees, for a period of ten years, for Rs. 50,000, which would have deprived the tenants of their rights in the jungle, affecting them seriously. This was more important as the settlement of the district of Manbhum had not yet been made and consequently the rights of the tenants in the jungles had not been recorded.[54]

It was on the basis of all this information regarding the indebtedness of the estate and the reasons mentioned above that the Commissioner, Chota Nagpur Division, also recommended its being taken under Government management under the provisions of the Encumbered Estates Act.[55] According to the statements of assets and liabilities of the estate prepared

[54]Letter No. 667 W/IX-1 dated 5th/11th May, 1915, as quoted in foot-note number 53.
[55]*Ibid.*

by the Deputy Commissioner, Manbhum, the annual income of the estate was estimated at Rs. 20,940 and expenditure at Rs. 6,710, leaving a surplus amount of Rs. 14,230 per year for the repayment of the debts. Maintenance allowance for the zamindar had not been taken into account in that scheme as the zamindar had got some *ghatwali* lands (lands held in consideration of police duties) yielding an annual income of about Rs. 6,000 in the district of Santal Parganas. As the Encumbered Estates Act was applicable only to the Chota Nagpur Division, his *ghatwali* property in the district of Santal Parganas, which was not situated in the Chota Nagpur Division, could not be taken over under the provisions of the Chota Nagpur Encumbered Estates Act. The zamindar was expected to meet his expenses out of the income of that property.[56] The zamindar of the estate, however, protested against the recommendations of the Deputy Commissioner, Manbhum, and the Commissioner, Chota Nagpur Division, to bring his estate under the protection of the Encumbered Estates Act. In the petition dated October 14, 1914, addressed to the Commissioner, Chota Nagpur Division, he stated that the debts of the estate at that time amounted to Rs. 14,380 only, and that he could repay them within a fortnight.[57] The Board of Revenue, after due consideration, came to the conclusion that a sufficiently strong case had not been made out for taking the estate under Government management against the will of the zamindar[58] and therefore no action was taken in the matter.

On August 31, 1920, the zamindar leased out surface and sub-soil rights in thirty-five villages, by a registered deed, for a period of 999 years, on a consideration of Rs. 1,30,000 as *salami* and an annual rent of Rs. 4,410, besides cess of Rs. 258-2-3, allowing a deduction of Rs. 100 on account of collection charges.[59]

In 1922, the zamindar obtained a loan of Rs. 20,000 under a registered usufructuary mortgage deed dated February 18, 1922. For the satisfaction of this debt with interest, he assigned thirteen villages, the annual income of which was Rs. 1,360-12-3 as rent and Rs. 89-14-9 as cesses, in all Rs. 1,450-11-0, for a period of twelve years. It was stipulated in the deed that the rent and cesses might be collected from the tenants and tenure-holders of the villages and appropriated towards the interest due at first and

[56]Letter No. 1365 W., dated November 25, 1914, from the Deputy Commissioner, Manbhum, to the Commissioner, Chota Nagpur Division (*vide* File No. 20-271 of 1915 of the Board of Revenue, Bihar and Orissa).

[57]*Vide* File No. 20-271 of 1915 of the Board of Revenue, Bihar and Orissa.

[58]Resolution dated June 5, 1915, of the Board of Revenue, Bihar and Orissa (*vide* File No. 20-271 of 1915 of the Board of Revenue, Bihar and Orissa).

[59]Statement of Facts of the General Manager, Wards and Encumbered Estates, Manbhum, in an Appeal filed in the Patna High Court against the order of the Sub-Judge, Dhanbad, in the Miscellaneous Case No. 106 of 1944 (*vide* File No. 20-618 of 1944 of the Board of Revenue, Bihar).

then to the principal debt. If at the end of the period of assignment, the zamindar was unable to pay off the balance, if any, the villages were to be continued to remain in the possession of the mortgagee till satisfaction of the loan amount advanced by him. On December 6, 1927, the mortgagee transferred his right, title and interest in those villages to the proprietor of a colliery by a deed of assignment. At the end of twelve years, it was found that a sum of Rs. 18,007-5-3 was due to the mortgagee and therefore it was decided to allow him to remain in possession of the villages assigned to him.[60]

Considering the speed with which the zamindar was giving *mukarrari* leases, it was no wonder that by the time of the survey and settlement operations in the district in 1918-1925, one hundred and seventy-nine entire villages (60.47 per cent) and eighty-six villages (29.05 per cent) in part out of a total number of two hundred and ninety-six villages of the estate had passed into the hands of tenure-holders (*vide* Table 5.1 given in the previous chapter). In terms of land, out of an area of 97,508.71 acres of the estate, 74,007.29 acres, that is 75.90 per cent, were under the control of tenure-holders at that time as will be seen from Table 5.2 given in the previous chapter.

On July 1, 1926, the minor son of the zamindar filed a petition through his mother saying that his father was drifting in the direction of ruin and therefore the estate be taken over by the Government under the protection of the Encumbered Estates Act.[61] The Additional Deputy Commissioner, Dhanbad, reported against the proposal and the Deputy Commissioner, Manbhum, agreed with him on the ground that it was doubtful if a satisfactory scheme could be made out for the management and for clearing off the debts within a reasonable time.[62]

On September 10, 1930, the minor son of the zamindar moved the Board of Revenue again for the protection of the estate stating that the estate was on the brink of ruin on account of mismanagement by the Dewan who had secured for himself a good deal of estate lands. He further stated that it had already run into debt to the extent of three lakhs of rupees which was increasing daily. Apprehending the future disaster of the estate, he

[60]Letter No. 706, dated April 2, 1935, from the General Manager, Wards and Encumbered Estates, Manbhum, to the Deputy Commissioner, Manbhum (*vide* File No. 20-314 of 1935 of the Board of Revenue, Bihar and Orissa).

[61]Petition dated July 1, 1926, from the minor son of the zamindar of the Tundi estate through his mother, as referred to in Letter No. 1633 R., dated August 12, 1926, from the Additional Deputy Commissioner, Dhanbad, to the Deputy Commissioner, Manbhum (*vide* File No. 20-349 of 1931 of the Board of Revenue, Bihar and Orissa).

[62]Order dated August 20, 1926, passed by the Deputy Commissioner, Manbhum, on the Additional Deputy Commissioner's above-mentioned Letter No, 1633 R., dated August 12, 1926 (*vide* File No. 20-349 of 1931 of the Board of Revenue, Bihar and Orissa).

prayed that it might be taken under Government management under the provisions of the Encumbered Estates Act immediately.[63]

The affairs of the estate continued to go from bad to worse so much so that on October 28, 1930, the estate was again advertised for sale in the Court of the Subordinate Judge of Dhanbad in the Execution Case No. 266 of 1930, under orders of the Calcutta High Court under the Suit No. 269 of 1928. The amount of the decree was Rs. 9,670 and the whole of the estate was to be put up for sale. Though the local officials had little sympathy with the improvident zamindar whose estate after years of trouble and labour on the part of the Government was restored to solvency to be again plunged into debt owing to mismanagement and extravagance on his part, yet the Deputy Commissioner, Manbhum, recommended the prohibition of sale of the immovable property of the estate or any portion thereof in execution of any decree or order of any civil or revenue court pending the issue of final orders on the petition of protection filed by the minor son of the zamindar.[64] The Commissioner, Chota Nagpur Division, issued the necessary order.[65]

During the course of a preliminary enquiry, ordered by the Deputy Commissioner, Manbhum, on October 14, 1930, and made by the General Manager, Wards and Encumbered Estates, Manbhum, about the financial position of the zamindar, it was found that the total amount of debt due by the zamindar amounted to about Rs. 3,50,000. The annual income of the estate was estimated by him at about Rs. 43,000 and expenditure at Rs. 24,213. As the important papers showing collections of miscellaneous rents, etc., were withheld by the zamindar and the Dewan, the General Manager, Wards and Encumbered Estates, had enquired regarding the assets and liabilities of the estate from some of the local money-lenders, shopkeepers and big tenants. The General Manager had also found that thirty-six villages had been given in *mukarrari* by the zamindar to the East India Coal Company. The Company in return had stipulated to pay one-and-a-half lakhs of rupees in liquidation of the debts of the zamindar. Those villages were subsequently transferred by the East India Coal Company to another person. Neither the East India Coal Company nor the latter had paid for all those debts. It was further reported by the General

[63]Petition dated September 10, 1930, from the minor son of the zamindar of the Tundi estate through his mother, enclosed with Letter No. 750 W/XIII-3, dated June 17, 1931, from the Officiating Commissioner, Chota Nagpur Division, to the Secretary, Board of Revenue, Bihar and Orissa (*vide* File No. 20-471 of 1931 of the Board of Revenue, Bihar and Orissa).

[64]Letter No. 1353 W., dated October 22, 1930, from the Deputy Commissioner, Manbhum, to the Commissioner, Chota Nagpur Division (*vide* File No. 20-349 of 1931 of the Board of Revenue, Bihar and Orissa).

[65]Order of the Commissioner, Chota Nagpur Division, dated October 28, 1930 (*vide* File No. 20-349 of 1931 of the Board of Revenue, Bihar and Orissa).

Manager that the Dewan of the estate was dead against the scheme of bringing the estate under Government management, and the popular idea was that it was he who was exploiting the resources of the estate.[66]

As a result of an enquiry conducted by the Deputy Collector, Dhanbad, about the financial status of the estate, which was found in chaotic condition, he recommended its protection on the ground that if the estate was allowed to disintegrate, its tenants would certainly suffer in the first instance. He thought that disintegration of the estate would undoubtedly be cause of unrest, and it was considered important at that time that all avoidable causes of unrest should be avoided. Another reason why it was recommended that the estate should be taken over under the protection of the Encumbered Estates Act was that there were very large areas of forest lands in the estate which had been neglected and were rapidly deteriorating. The jungle was being wasted mercilessly and big trees had become rare. The reasons for this wanton destruction were obvious. The estate was in grave financial difficulties. The employees of the estate rarely got their salary. They compensated themselves by other means. They sold the jungle and issued permits for extraction of wood from the jungle without any authority; there was nobody to detect that. Even if they had taken permission for extraction of a single tree, they allowed extraction for many trees. Thus the staff was in this way responsible for the damage done in the *rakh* (reserved) forest. The tenants, especially the non-Santals, whose rights over the jungle were more limited, were utilising that opportunity. The Santal tenants were also not behind in the job. They were exceeding their rights. The zamindar did not want to fight out cases on that account in the courts as he had no money for that and therefore had given it up as a lost game. This was the condition of the forest which was directly under the control of the estate. But in villages under the control of tenure-holders, it was just the opposite and the forest was reported to have been kept under very strict supervision. The proper care of forest lands appeared to be of more than local interest. The Deputy Collector also reported that the zamindar was a nonentity and the entire management of the estate was in the hands of the Dewan.[67]

The enquiry conducted by the Deputy Commissioner, Manbhum, showed that the estate had a considerable income from the forest. Though he was

[66]Report of the General Manager, Wards and Encumbered Estates, Manbhum, dated November 10, 1930, enclosed with Letter No. 1452 W., dated November 14, 1930, from the Deputy Commissioner, Manbhum, to the Commissioner, Chota Nagpur Division (*vide* File No. 20-349 of 1931 of the Board of Revenue, Bihar and Orissa).

[67]Report of the Deputy Collector, Dhanbad, dated December 29, 1930, enclosed with Letter No. 50 R., dated January 7, 1931, from the Additional Deputy Commissioner, Dhanbad, to the Deputy Commissioner, Manbhum (*vide* File No. 20-349 of 1931 of the Board of Revenue, Bihar and Orissa).

handicapped in his enquiry by the disinclination of the zamindar and his Dewan to disclose the correct assets and liabilities of the estate, he prepared a fairly detailed account of the financial position of the estate. The Deputy Commissioner also was of the opinion that the zamindar was entirely in the hands of his Dewan. During the course of his investigations, it came to light that the Dewan had settled bamboo and wood in the hills with his son. In the statement of the scheme of management of the estate prepared by the Deputy Commissioner, its annual income was shown as Rs. 48,330 with an annual expenditure of Rs. 19,080, leaving a yearly surplus of Rs. 29,250 for repayment of debts amounting to Rs. 3,25,826 (Rs. 2,58,692 principal and Rs. 67,134 interest) in the course of seventeen years, that is by the year 1947-1948 from the year 1931-1932. Though according to the Rules framed under the Chota Nagpur Encumbered Estates Act it was required that the debts should be liquidated within fifteen years, it was thought impossible to fulfil that condition in this case.[68]

When the creditors of the estate came to know that a proposal had been set on foot to bring the estate under protection of the Encumbered Estates Act, eight of them submitted an application to the Commissioner, Chota Nagpur Division, on March 2, 1931, stating that no scheme was possible to be formulated to enable the estate to liquidate its liabilities within the statutory period as the income, after defraying the Government dues, collection and other charges, would not be even sufficient for meeting the interest payable annually to the creditors. They further stated that if under such circumstances the estate was taken over by the Government under its management, hundreds of bona-fide creditors would be ruined forever, as it would be quite uncertain when the estate would be able to pay up the debts, if at all. According to their information the annual income of the estate from all sources did not exceed Rs. 31,000 and the debts amounted to nearly Rs. 3,20,000. Thus they vehemently opposed the idea of bringing the estate under the protection of the Encumbered Estates Act.[69] On April 4, 1931, the above-mentioned creditors filed another application stating several facts in support of their contention and challenging the correctness of the account submitted on behalf of the zamindar, and praying once again that the estate should not be brought under Government management.[70] In spite of strong opposition from the Dewan, who was obviously making money out of it and therefore whose interest was antagonistic to the taking over of the estate under Government management, and by the creditors, who expected to make more by selling up the

[68]Letter No. 552 W., dated 8th/9th May, 1931, from the Deputy Commissioner, Manbhum, to the Commissioner, Chota Nagpur Division (*vide* File No. 20-349 of 1931 of the Board of Revenue, Bihar and Orissa).

[69]*Vide* File No. 20-349 of 1931 of the Board of Revenue, Bihar and Orissa.

[70]*Ibid.*

estate than by getting their just dues paid off by the court, the Board of Revenue recommended the taking over of the estate under the protection of the Encumbered Estates Act[71] in accordance with the recommendation of the Commissioner of the Chota Nagpur Division who had stated in his letter that he had received an application from the zamindar also praying for the management of his estate. The Commissioner was given to understand that the zamindar was too much afraid of his Dewan to file such an application openly.[72] Though the Government was loth to extend any helping hand to such zamindars, it sanctioned the proposal as a special case.[73] The estate was taken under the protection of the Encumbered Estates Act on September 16, 1931.[74]

Though at the time of assumption of the charge in 1931 the period required for liquidation of debts was estimated at seventeen years, by 1934-1935, while preparing the final scheme, it appeared to the authorities that it would not be possible to wipe out the debts by that time. In the preliminary scheme, the maintenance allowance of the zamindar and his family was provided at Rs. 6,865 out of which Rs. 2, 365 were payable from the estate's fund and the balance of Rs. 4,500 was to have been met out of the income of the *ghatwali* tenure of the zamindar which had not been taken under management as it lay outside the Chota Nagpur Division. It was later found that that tenure was also saddled with debts and at the instance of one of the creditors had been placed under the Manager as Receiver. The liabilities on that property were not known. On account of those liabilities and the appointment of the Receiver the arrangement for payment of maintenance allowance to the zamindar was upset. Secondly, in the preliminary scheme certain incomes were also found to be overestimated. As a result of these factors the debt repayment had fallen behind the schedule which necessitated the revision of the scheme.[75] The final scheme showed the annual income of the estate as Rs. 34,184. A

[71]Letter No. 20-349-9, dated July 22, 1931, from the Secretary, Board of Revenue, Bihar and Orissa, to the Secretary to the Government of Bihar and Orissa, Revenue Department (*vide* File No. 20-349 of 1931 of the Board of Revenue, Bihar and Orissa).

[72]Letter No. 750 W/XIII-3, dated June 17, 1931, from the Officiating Commissioner, Chota Nagpur Division, to the Secretary, Board of Revenue, Bihar and Orissa (*vide* File No. 20-349 of 1931 of the Board of Revenue, Bihar and Orissa).

[73]Letter No. 998-R.R./IW-50, dated August 21, 1931, from the Officiating Secretary to the Government of Bihar and Orissa, Revenue Department, to the Secretary, Board of Revenue, Bihar and Orissa (*vide* File No. 20-349 of 1931 of the Board of Revenue, Bihar and Orissa).

[74]The Bihar and Orissa Government Notification dated September 9, 1931, issued by the Commissioner, Chota Nagpur Division, published in *The Bihar and Orissa Gazette* dated September 16, 1931, Part I, p. 408, enclosed with Letter No. 1271 W/XIII-3, dated September 28, 1931, from the Personal Assistant to the Commissioner, Chota Nagpur Division, to the Secretary, Board of Revenue, Bihar and Orissa (*vide* File No. 20-349 of 1931 of the Board of Revenue, Bihar and Orissa).

[75]Letter No. 239 W/XVIII-2, dated March 6, 1935, from the Personal Assistant

provision of Rs. 18,184 was made for the annual expenditure which left a surplus of Rs. 16,000 a year for repayment of debts. The annual interest due to the creditors was estimated at Rs. 12,560. The total debts, as shown in the final scheme, amounted to Rs. 3,27,183-7-5½ (Rs. 2,30,916 principal and Rs. 96,267-7-5½ interest) which, it was estimated, would take twenty-five years more to liquidate. Thus the total period of management was reckoned as twenty-nine years from the year of assumption of the charge. But the question then was whether the estate should be retained under management or it should be released. The crux of the matter was whether there was any real advantage in continuing to manage an estate of that kind. It was obvious that the amount of trouble which had been taken in the past by the Government to confer solvency on the estate had not produced results at all commensurate with that. The officers of the Government had worked for years to reduce the affairs of the estate to order, and within a short period of its release from Government it was as hopelessly encumbered as ever. It was said that even the period of twenty-nine years was optimistic since the scheme had been prepared on the basis of income for normal years only without taking into account the fall in income during abnormal years of drought, etc., and extraordinary charges. The Commissioner, Chota Nagpur Division, however, recommended the retention of the estate under protection,[76] the Board of Revenue approved the proposal[77] and the Government accepted the Board's approval.[78] According to that scheme the estate was to be released in 1959-1960.

In 1942-1943, however, the scheme was again revised. The revised scheme was to run for a period of nineteen years from 1942-1943 and the estate was to be released in 1960-1961. Thus the management was extended by one year. The necessity for revising the scheme had arisen on account of the rent reduction operations. In the revised scheme, the annual income was shown as Rs. 36,050 with an expenditure of Rs. 18,476, leaving an annual surplus of Rs. 17,574 for the repayment of debts, then amount-

to the Commissioner, Chota Nagpur Division, to the Deputy Commissioner, Manbhum (*vide* File No. 20-314 of 1935 of the Board of Revenue, Bihar and Orissa).

[76]Final Scheme of the Tundi Encumbered Estate enclosed with Letter No. 787 W/XVII-2, dated June 26, 1935, from the Commissioner, Chota Nagpur Division, to the Secretary, Board of Revenue, Bihar and Orissa (*vide* File No. 20-314 of 1935 of the Board of Revenue, Bihar and Orissa).

[77]Memo No. 20-314-2, dated July 13, 1935, from the Secretary, Board of Revenue, Bihar and Orissa, to the Secretary to the Government of Bihar and Orissa, Revenue Department (*vide* File No. 20-314 of 1935 of the Board of Revenue, Bihar and Orissa).

[78]Letter No. 6048R/IW-74, dated August 12, 1935, from the Assistant Secretary to the Government, Revenue Department, to the Secretary, Board of Revenue, Bihar and Orissa (*vide* File No. 20-314 of 1935 of the Board of Revenue, Bihar and Orissa).

ing to Rs. 2,97,041, in nineteen years.[79] The scheme was recommended by the Commissioner[80] and sanctioned by the Board of Revenue.[81]

At the close of the year 1944-1945, the debt repayment fell behind the scheme by Rs. 24,826 and the annual expenditure also increased considerably. Consequently the scheme was further revised. In the revised scheme the annual income and expenditure were shown as Rs. 46,316 and Rs. 28,913 respectively, leaving a surplus of Rs. 17,403 per annum for repaying the debts amounting to Rs. 2,22,778 in the course of fifteen years from 1946-1947. Thus there was no increase in the period of management.[82] Though this scheme also was recommended by the Commissioner, Chota Nagpur Division,[83] and approved by the Board of Revenue,[84] a letter from the Personal Assistant to the Commissioner of the Chota Nagpur Division to the Secretary, Board of Revenue, shows that the revised scheme of the estate which began from 1946-1947 was also not working according to the provisions made therein as the debt repayment had fallen short of the provisions in the scheme by Rs. 15,478 in the first year itself of the scheme.[85] Whatever may have been the reasons for the short payment in that year, it had already been anticipated by the local officials that the scheme of the Tundi estate was "going to be a permanent affair".[86] The Tundi estate vested in the Government as an encumbered estate at the time of the abolition of the zamindari system in 1954.

[79]Revised Scheme of the Tundi Encumbered Estate enclosed with Letter No. 874, dated August 31, 1942, from the General Manager, Wards and Encumbered Estates, Manbhum, to the Deputy Commissioner, Manbhum (*vide* File No. 20-493 of 1943 of the Board of Revenue, Bihar).

[80]Letter No. 134 W., dated January 3, 1943, from the Commissioner, Chota Nagpur Division, to the Secretary, Board of Revenue, Bihar (*vide* File No. 20-493 of 1943 of the Board of Revenue, Bihar).

[81]Letter No. 20-493-2, dated February 22, 1943, from the Under Secretary, Board of Revenue, Bihar, to the Commissioner, Chota Nagpur Division (*vide* File No. 20-493 of 1943 of the Board of Revenue, Bihar).

[82]Revised Scheme of the Tundi Encumbered Estate enclosed with Letter No. 128, dated 1st/2nd January, 1947, from the General Manager, Wards and Encumbered Estates, Manbhum, to the Deputy Commissioner, Manbhum (*vide* File No. 20-779 of 1947 of the Board of Revenue, Bihar).

[83]Letter No. 634 W., dated May 26, 1947, from the Personal Assistant to the Commissioner, Chota Nagpur Division, to the Secretary, Board of Revenue, Bihar (*vide* File No. 20-779 of 1947 of the Board of Revenue, Bihar).

[84]Letter No. 20-779/47-4, dated April 5, 1948, from the Secretary, Board of Revenue, Bihar, to the Commissioner, Chota Nagpur Division (*vide* File No. 20-779 of 1947 of the Board of Revenue, Bihar).

[85]Letter No. 136 W., dated February 3, 1948, from the Personal Assistant to the Commissioner, Chota Nagpur Division, to the Secretary, Board of Revenue, Bihar (*vide* File No. 20-779 of 1947 of the Board of Revenue, Bihar).

[86]Letter No. 18 W., dated January 7, 1948, from the Senior Deputy Collector Incharge, Manbhum, to the Personal Assistant to the Commissioner, Chota Nagpur Division (*vide* File No. 20-779 of 1947 of the Board of Revenue, Bihar).

Conclusion

The study of the financial position of the two zamindaris, one coal and the other non-coal, shows that the so called baronial houses had come to rack and ruin at the hands of irresponsible zamindars. The more valuable properties a zamindar had, the more indebted he was because he could borrow more on the security of his property. As their extinction was not viewed with unconcern and as their retention was considered almost a sacred and pious duty and also a vital necessity because it was supposed to have a wider significance and importance to the territory as a whole, a unique piece of legislation, namely the Chota Nagpur Encumbered Estates Act, was enacted to protect those zamindars as monuments are protected.

The intention of enacting the Chota Nagpur Encumbered Estates Act was to prevent the supplanting of the old feudal landlords by new immigrants who were presumably ignorant of the proper methods of handling aboriginal tenants for whose protection there was no suitable tenancy Act, no record-of-rights and about whose customs the courts did not know much. The conditions had changed later on with the enactment of the Chota Nagpur Tenancy Act, 1908, and the completion of the record-of-rights which placed the aboriginals in a very strong position. They were no longer at the mercy of their landlords. But the operation of the Act continued unabated, if not more vigorously. Secondly, with an astonishing degree of sub-infeudation promoted by this system, it is doubtful whether the interests of the tenants at the lowest level would have been seriously affected merely because of the change in the superior landlords. Thus it shows that the existence of the Act was more to save the spendthrifts, who had embarked on a course of extravagance which was likely to lead to the break-up of their estates, than to preserve the interests of the tenantry. Though the Government officials always questioned the propriety of extending protection to the improvident zamindars and did not sympathise with such extravagants, yet they thought that so long as the Act existed some measures of protection ought to be afforded to the children of those zamindars, who, through no fault of their own, found that they were being deprived of their ancestral property owing to the deliberate policy of waste entered into by their fathers. But the result of this study reveals that in endeavouring to save the estates from dissipation they were attempting the labour of a sisyphus. The estate which had been brought up to a solvent condition, when made over to a reckless zamindar, at once began to run down-hill again and in a short time was qualified for a second period of management. The net result of the Act was to discourage influx of finances for development and that put a stop to all progress and exploiting of potentialities.

7

Henner Papendieck

British Managing Agencies in the Indian Coalfield

1. *Introduction*

THE following analysis deals with the effects of a colonial type of capital organisation and control on the development of one of the important sectors of the Indian economy, the coal mining industry during its most rapid expansion. The period from 1890 to 1920 was chosen as the phase during which the structure of Indian coal mining developed those features which led to stagnation and hindered a modern development. From its beginnings in 1820 until the 1880s the coal mining industry of Bengal grew in accordance with the needs of various departments (Military, the Mint, etc.) of the East India Company. The industry, centered in the large Raniganj-field, not far from Calcutta, grew gradually with the help of government contracts, guaranteed purchases and direct subsidies for the establishment of an infrastructure for distributing coal to sea-going harbours and bunkers of steamer stations along the Ganges.[1] Two phases of more rapid growth, the first in the 1830s and 1840s (due to the growing demand of the steam-boats) and the second from the 1860s to 1880s (due to the expansion of

[1]The Marine Board Proceedings (from 1834 onwards), the General Marine Department Proceedings (from 1840 onwards), and the Bengal Steam Proceedings (all with the India Office Records, London) contain much information on the various attempts of the government to develop an Indian coal mining industry and to set up a network for the distribution of bunker coal.

railways in India), led to a volume of production between 1 and 1-1/2 million tons p.a. in the decade from 1880 to 1890.[2] This amount of coal was mainly raised by a small number of pioneer companies (all of them established with Rupee-capital), enjoying stable and rarely exorbitant profits. Although the capital invested in these companies did not stem from Britain but was raised in India, part of their profits was transferred to Britain, as investors went back to their home country after leaving the service of the East India Company or that of one of the Government departments. Reinvestment of profits was low.[3]

These pioneer companies were independent and competed with each other for government and railway tenders. Attempts to monopolize the coalfields, which had been made by the Bengal Coal Company from its start onwards, were unsuccessful. Nevertheless, it was this company that dominated the development of the Raniganj coalfield until around 1890.[4]

These companies found themselves in surroundings hardly apt to control the behaviour of a modern capitalist enterprise. There were no mining laws or government controls of mining operations. There was even uncertainty about the ownership of mining rights, royalties etc.[5] The zamindars, from whom the companies obtained mining concessions, surface and underground leases, lacked all knowledge of coal mining, economic working of seams, possible damages resulting from mining operations, as well as the profit opportunities in coal mining.

The companies decided without any interference which seam they would work on and which means to employ in raising the coal. They acquired extensive leases and built up large reserves of mining land for future exploit-

[2]For figures on the production up to 1890 cf. Oldham, T: Coal Resources and Production of India. Calcutta, March 1867, p. 26, and: Commercial Intelligence Department, Government of India, Note on the Production and Consumption of Coal in India up to the year 1906. Calcutta 1907.

[3]In some companies the directors were denied the right to invest more than a very small percentage of the company's profits without the prior consent of the shareholders. The Bengal Coal Company in its earlier years is one of the well-known examples for this restrictive policy.

[4]In the early 1880 the share capital of the Bengal Coal Company amounted to 2.2 million Rupees, almost half of all capital invested into coal mining companies in Bengal. Cf. Returns on Joint Stock Companies. Bengal Registration Dept., 1879, 1880, 1880/81 and 1882/83 (West Bengal State Archive, Calcutta).

[5]The issue was seriously taken up by the Government of Bengal in 1872. In June of that year, H.L. Dampier, Secretary, Government of Bengal wrote to the Secretary, Govt. of India: "The fact is, that till the last few months, nothing has been expressly reserved for Government: on the contrary, as will be seen from the Board's resume, all discussions have generally ended in the abandonment of claims of the public and the concessions of the rights, claims, or demands of individuals." (West Bengal State Archive, Calcutta: Govt. of India, Dept. of Agriculture, Revenue and Commerce. Minerals and Geological Survey, August 1872, Proc. No. 37).

ation. In their organisation of the labour force they took over many features of the feudal system prevailing in Bengal. They established their own zamindari-departments and took over the role of the zamindars in sub-leasing land to their miners, a system under which the miner owed his labour as well as rent to the company. In this way there grew a labour force of miners settled on the land surrounding the mines, working partly as coal miners, partly as small-scale farmers.

This state of relative equilibrium (in no way a peaceful equilibrium) was shattered by the rapid expansion of coal mining in Bengal in the three decades between 1890 and 1920. A new structure developed, absorbing some features of the old system, but pushing aside most of the hitherto prevailing characteristics.

For the coal mining companies the parameters for decision-making changed fundamentally within this relatively short period of time. The newly established companies, in a position of newcomers vis-a-vis the pioneer companies, had to act in a totally different setting to the one the older companies had found. Labour migration set in on a scale which had previously only been observed in labour migration to Calcutta or the tea estates of Assam. In the search for better qualities of coal, shafts had to be sunk to new depths, which required more sophisticated technical equipment for the shafts as well as installations, like ventilation, pumping equipment etc. The question of mechanisation cropped up, both in hewing the coal and in raising it to the surface. It became doubtful if steam power would be adequate for these new adaptations, electrification had to be thought of, and this needed relatively large investments both for power houses and for the net of installations to supply electric current to the pits.[6]

The transport of the bulk commodity coal posed new problems. The distribution of coal required new and larger stocks at various places all over the subcontinent. Capital requirements for investments in construction and in stock rose rapidly. Parallel to this development the total freedom of the companies' decision-making came to an end. Raising contracts and leases began to contain specifications about the quantity and the quality of coal to be raised. Royalties were tied to the actual output of coal, whereas many rents for leases had previously been tied to the area of surface land covered by the lease. Buying up large stretches of reserve land became almost impossible and new cost-considerations had to be taken into account. The old strategy of delaying the mining operations over years was penalised. The turn of the century saw the beginnings of government regulations for the coal mining industry, introducing some measures necessitating

[6]The discussions which were taking place in the industry are reflected in the publication: *Transactions of the Mining, Geological and Metallurgical Institute of India*, Calcutta 1907 ff.

investments to meet safety standards.[7] The European experience of industrialisation would indicate that competition should have grown fiercer as more and more companies entered coalfields, the input of capital should have become larger as compared to the input of labour, and companies should have been forced to rationalise and mechanise in order to raise productivity. In the process of an increasing division of labour, one would expect a skilled labour force to emerge, permanently employed, receiving higher wages and consequently an improvement of the miners' standard of living. Such developments might have marked the steps of the coal mining industry into a period of sustained growth under the conditions of capitalist industrialisation. However, none of these developments took place in the coalfield of Bengal.

The stunted growth of coal mining in India running into stagnation after a relatively short period of rapid expansion could easily be attributed to one factor: the lack of an Indian iron industry. Undoubtedly, this factor is of great importance. But it does not explain why the growth was stunted. In spite of the lack of an iron industry there was a rapidly increasing demand for coal in India and in the export markets around India. And this demand in export markets like Colombo, Singapore, etc. was geared towards high quality coal. So the incentives for growth did not only stem from the total substitution of coal imports, first in East India, later along the western coast, but was paralleled to a large extent by rising export opportunities. The only limits on these export markets was the competition from other Asian and African countries: Japan, Australia and South Africa, all of which had much longer sea routes to these markets. In other words: the conditions for substantial growth of the coal mining sector in Bengal did exist.

None of the above-mentioned assumptions correspond to the development of the coal mining industry of India until the end of World War I. Due to wasteful exploitation of coal seams, large quantities of coal were lost; the equipment of coal mines was totally outdated, rundown and unsafe, hardly any mines were mechanised and only a few exceptional coal companies had begun sinking deep shafts. The average size of coal mines had not risen by any considerable degree since the turn of the century, nor had the output per mine or the productivity (it remained almost constant from 1906 to 1920). Rationalisation had not taken place, chances for large scale exports had been missed. There was no skilled and specialised work force of coal miners, real wages of coal miners had not risen but fallen, and the raising capacity of the Indian coal mining industry by the end of World War I was not higher than in the first decade of the 20th century.

[7]The question of a "legislation for the working of coal mines" had been discussed over decades. Twice before, in 1870 and 1880 it had been turned down. Finally in 1895 the office of an Inspector of Coal Mines was created.

Certainly the effects of the war economy, i.e. the controls imposed on distribution and prices of coal in the years 1916-1918 should not be underestimated. They certainly did not help to increase production of coal but had effects to the contrary. The controls, it seems, made the management of the coal companies acutely aware that profits could be rapidly increased by keeping down production in a situation of rising demand.

Apart from the disruption of the growth pattern by government controls in the war years there was an amazing continuity throughout the phase of rapid development in the Indian coal mining industry: a continuity of extremely large profits syphoned off by a number of fairly large coal companies, around them a set of smaller companies, and an increasing influence of the peculiar managing agency system on the industry, a high degree of centralisation of capital, but a lack of any corresponding concentration of production in large units and of any economies of scale. The influence of the managing agents was by no means confined to the coal mining sector. It comprised the whole of the industrial sector of Bengal as well as the tea plantations, large parts of the tertiary sector—banking, trade, traffic networks, steamlines, railway lines etc., and in some cases even zamindari estates.

It is the obvious discrepancy of low productivity, low standards of technology and mechanisation, declining standards of living and working for the miners on the one hand and the amazing prosperity of the coal companies on the other hand, which will be the subject of the following analysis. As the penetrating power of the managing agency system was clearly demonstrated by the fact that after 1908 there was not a single public company in coal mining in Bengal, which was not controlled by one managing agent or another, we shall concentrate on analysing this system in trying to find the answer to the problem of stunted growth.

In the attempt to locate the specific impact of this system on the coal mining industry we have to deal with the inner mechanism of the system itself. The material at hand originates from the coal departments of these business houses. The coal department played an important role in some of the managing agents' organisations, it certainly was important in the two largest European establishments in East India, the firms of Andrew Yule & Co and Bird & Co. Nevertheless, coal interests constituted the periphery rather than the centre of the network of financial transactions of these houses. Their centres were the firms of the managing agents themselves, organised in terms of partnership contracts and linked with their own banks; in the case of Andrew Yule & Co. the Calcutta Bank and later the Mercantile Bank of India Ltd.[8]

[8]From the books of the Katras-Jherriah Coal Co. we can reconstruct that the merger of the Calcutta Bank with the Mercantile Bank of India, an international bank with branches all over South and Southeast Asia was perfect in 1906.

After a short quantitative introduction into the main features of the development of coal mining in Bengal between 1890 and 1920 this study will contain a short account of the inner working of the managing agency system, vis-a-vis capital investors as well as the companies they controlled. It is quite obvious from the start that the managing agents tended to eliminate the independence of companies controlled by them on a horizontal level (by tying companies of the same industrial sector together) and tended to level out demand and output capacities of their companies on a vertical level. The existence of interdependencies among companies of one group is quite obvious. Yet the nature and the extent of these interdependencies is still a puzzle. So far there has not been a thorough analysis of "all the various ways of mutual accommodation" at work within managing agency houses, about which well informed people in Calcutta talked about at the turn of the century.[9] Previous studies of the managing agency system have not attempted to (or their sources did not permit them to) bring to light the operational structure of the system and to demonstrate its effects on industrial development in Bengal.[10]

This operational structure, however, is of special importance and interest. We know that the managing agencies started out by administering capital of third parties and it took them no longer than a few decades to become holding companies, looking after their own capital interests. How did this come about? How did they come to control the majority of shares within the key companies? There is evidence to suggest that this process of consolidation of their own interests did not take place in the 1920s and 1930s, but in the period of our study. One obvious sign for it seems to be the amalgamation of the whole coal department of Bird & Co in the Burakur Coal Company in 1919.[11]

As long as we do not have access to data stemming from the central units within the managing agencies it is impossible to show the flow of capital on an international scale, the repatriation of profits to Britain or the invest-

[9]The chairman of the Bengal Coal Co, explained in 1900: "It is all the more important that our company should devote itself to the development of the export trade, as the local trade is passing more and more every day into the hands of various commercial firms in Calcutta. Many of these firms act at one and the same time as agents for coal companies and for mills, tea gardens etc., and in some cases for steamer companies. A steamer company, for example, will be induced to take coal from that firm alone, which will undertake to ship jute and tea by this company's steamers, through all the many moving influences of reciprocal accommodation". Capital, Calcutta, 24.5. 1900, p. 767.

[10]Cf. Tyson, Geoffrey William: *Managing Agency: A system of business organisation*. Calcutta, 1961: Basu, S.K.: *The Managing Agency System in Prospect and Retrospect*, Calcutta, World Press 1958; *Report of the Managing Agency Enquiry Committee*, Government of India, Ministry of Law, Dept. of Company Affairs, 1966: et al.

[11]Cf. Harrison, Godfrey: *Bird and Company of Calcutta, A History produced to mark the Firm's Centenary, 1864-1964*. Private Publication, Calcutta, 1964.

ment of capital accumulated in India in other parts of the world. But data from the periphery of this system will certainly give valuable hints on the working of the system and will enable us to draw some conclusions on the effects of the system.

The managing agency system in Bengal was the specific organisational form of British imperialism in India. So the question of the "private drain" as a counterpart to the "public drain" is pre-eminent in our context. Did British-controlled capital interests reinvest into the industries they controlled and thus contribute to the overall capital accumulation needed in the process of industrialisation or did they withdraw the profits from this sector, prevented capital accumulation, and thus hindered the process of industrial development?

The present study tries to answer these questions on the basis of data from the archives of the coal department of Andrew Yule & Co, the largest and most profitable managing agency house in Bengal. This house was not typical for all managing agents in Bengal in the sense of statistical averages. It is not accidental that it surpassed most of the others. But the power of penetration, which this system so evidently demonstrated, might come into sharper focus where it was most successful.

What is the material basis for reconstructing the "behaviour" of these companies? Our data stem from the journals (accumulated figures for 6 months), current cash books (daily entries) of various companies, in some cases their ledgers and in a few their sales registers.[12] From them we learn nothing about the entrepreneurial intentions, aims and strategies, expectations, fears etc. From them we only reconstruct the transactions and operations reflected in their account books. One could argue that this procedure resembles the attempt of establishing the nature of a fish by counting its bones. That is partly true. The reduced reality of accountancy however has one advantage: it focusses our attention on the structure of actual business deals, the pattern which asserted itself successfully, and reveals the structure behind it.

One note on the limitations of the data is necessary. They are expressed in monetary terms, which has an obvious advantage: we can easily compare all available data. The disadvantage is equally obvious: very often we would like to know more about the physical equivalents of the monetary expressions: the quantities of coal raised rather than the income from sales, the number of working hours or days rather than the wages paid out, the number of workers recruited rather than the cost of recruiting etc. Never-

[12]The author's dissertation contains a detailed list of the books which are available in the archive. Colin Simmons published an article in *Bengal Past and Present*, Vol. XCI, Calcutta 1972, Part II, entitled: Towards a Primary Source Bibliography for the Study of the Indian Coal Mining Industry. The list was made according to former transfer lists. Simmons obviously did not look into the archives themselves. He would have realised that the series are largely incomplete.

theless, it were the monetary terms which guided the investors and provided the basis for the managing agents' decisions. We shall have to follow this logic.

In the third and main part of the present study we will look into a limited number of aspects of the system. A larger number of aspects have been dealt with in the present author's dissertation. The technology of raising coal will not come into the picture, nor will living and working conditions of the miners. The first part of this section will concentrate on aspects of horizontal integration. How large was the volume of transactions between the coal companies under the same control? Do we find mutual shareholdings, credits etc.? Which were the effects of mutual ties? Did they lead to leveling off profits of companies under the same control, as Lokanathan suggested in his study of the managing agency system,[13] or did they work to the contrary? Secondly the question of vertical integration will be dealt with, the structure of sales. How much was sold to companies of the same group? Prices, of course, play a big role in this context. Pricing is the simple and effective means of transferring profits within large industrial groups.

How large were the profits? Was there a pattern of profit margins within the group? And finally how were these profits used, which part was reinvested, which part was used for accumulating reserves in various ways and which part was paid out in dividends?

The pattern of investments will give an indication about the type of development which took place under the control of the managing agency system in Indian coal mining. The answers to these questions might enable us to verify or contradict the inconsistent speculations about the effects of the managing agency system, which have hitherto made it difficult to assess its role as one of the key characteristics of British capital control in India.

2. *Sectoral Analysis*

The coalfields of India, stretching from the Raniganj field, some 100 miles west of Howrah, via Jharia, Bokaro Ramgarh, Karanpura, Auranga and Daltonganj deep into central India, belong to the largest known coal resources of the world. Of this chain of coalfields the largest lie within the former Presidency of Bengal, the present states of West Bengal, and Bihar.

The five most important coalfields of Bengal were opened in according to their distance from Howrah. Karanpura and Bokaro did not play any role before 1918 except for purposes of speculation. In our context, there-

[13]Lokanathan, P.S.: *Industrial Organisation in India* (LSF, Studies in Economics and Commerce No. 4), London 1935, Reprint Kraus 1970.

fore, we are dealing with three coalfields—Raniganj, with a surface area of roughly 1,300 square km and seams varying between 3 and 6 m in thickness (total volume estimated at 510 million tons of first class steam coal), Jharia, almost adjoining to Raniganj in the west, covering a surface area of about 400 square km, its seams reaching a thickness of up to 30 m (total volume estimated around 480 million tons), and the much smaller Giridih-field in the north of both. Throughout the nineteenth century the Jharia field had been connected with the network of the East Indian Railway, in 1894 its production grew rapidly and surpassed that of the Raniganj-field by 1907.[14]

2.1 *Conditions up to 1890*

In the introduction the term "relative equilibrium" was used to characterise the state of affairs which prevailed in Indian coal mining before the production of coal started to grow rapidly in the 1890s. Studies dealing with the structure of coal mining in India are scarce, even more so for the 19th century. Therefore, it seems necessary to sketch the main features of the development up to 1890.

The needs of the military department of the East India Company for coal (requirements for their arsenals) led to the first systematic exploration of coal resources in India from 1808 onwards. It was one of the results of these explorations that the first coal mine was opened in 1817 in Raniganj. It was soon joined by a number of competitors in the 1820s. The requirements of the mint and the military department were stimulants for further explorations and the search for higher qualities of coal, paralleled by a growing demand when steamboats were introduced on the Ganges in the 1840s. In order to supply the steamboats with coal along the steamer stations up to Allahabad a wide distribution system had to be set up, the cost of which was borne mainly by departments of the East India Company.

When this distribution network was later offered to the coal companies it provided the base for the even wider distribution system required with the arrival of the railways.[15] The railway line linking the coalfield of Raniganj with Howrah in 1854, for obvious reasons the first one to be built in Bengal, started a new era of development. The railways provided a larger and very soon the predominant demand for coal, which started to increase rapidly only from the late 1860s onwards (from 1859 until 1866 coal production remained almost constant at a level of less than 400.000

[14]The Jharia coalfield remained the largest coal producing area of India throughout our period, this coalfield reached the highest growth rates but also bore the large fluctuations between booms and depressions in the Indian coal mining industry.

[15]The companies had to set up large depots at various points. The largest depots were of course still kept to store bunker coal in Bengal.

tons p.a.).[16] This rising demand led the East Indian Railway to invest on a large scale into its own railway collieries, which were raising roughly a quarter of the total Indian coal production by 1880. Only after successfully eliminating the competition from this side (through legal proceedings against this "unfair competition") was the road for more or less untroubled growth of the private coal companies free.[17]

It was, however, not only the railways' demand for coal, that made a big change to the development of Indian coal mining but their facility to guarantee a regular coal supply independent of the season. Deliveries of coal to Calcutta had previously been confined to the rainy season, two or three months of the year, during which the Damodar river carried enough water to allow river transport in country boats.[18] The Bengal Coal Company, established in 1834 by the combination of three hitherto rivalling groups entertaining coal mines in Raniganj,[19] dominated the development of the Raniganj coalfield without any major threat of competition well into the 1860s. This company remained the largest private coal company in India until nationalisation in 1973.

It was joined in the following decades from 1860 to 1890 by four other public companies, the New Beerbhoom Coal Company (established in 1861 and registered as a Ltd. Co., in 1873), the Equitable Coal Company (1864), the Raneegunge Coal Association (1873) and the Burrakar Coal Co (1875), their accumulated capital (paid in by 1877) equaling that of the Bengal Coal Company, 2.2 mill. Rupees.[20] All of them worked in the Raniganj field, the Giridih-field was almost exclusively the dominion of the East Indian Railway Co.

These five pioneer companies raised about 80% of the private sector's coal production by 1890. They had been able to organise their production without breaking up the feudal zamindari system which they found in their

[16]Cf. Oldham, Thomas: *The Coal Resources and Production of India*. Being Return called for by the Right Honourable, The Secretary of State for India. Calcutta 1867.

[17]The struggle between the coal companies and the EIR came to an end when the coal companies addressed a petition to the Government of India in April 1881 in which they stressed "having been acquired at the expense of the general taxation of the Empire, it is unbecoming, or unseemly on this score, as well as on account of the illegality of the Act, that the sale of coal by the company (the EIR) . . . should be continued. "The Deputy Secretary of Public Railway Construction, Govt. of India, assured the Bengal Chamber of Commerce on 16th August, 1881 that "all danger of competition will be avoid". (B.C.C.-archives)

[18]Much of the early correspondence between the Mining Engineers of the Bengal Coal Company and the head office in Calcutta dealt with the rather dangerous transport of coal down the Damodar river during the rainy season.

[19]The main partners were Carr, Tagore & Co. Cf. contract between Carr, Tagore and Company and the owners of Naraincoyry Colliery of 1843 (B.C.C.-archive).

[20]Cf. *Return of Joint Stock Companies in Bengal for 1877*. Bengal Revenue Proceedings, Registration Dept., A 5-7, March 1878, Statement IV, (West Bengal State Archive).

rural surroundings. They recruited and controlled the labour force which they needed to raise the necessary 1.5 million tons of coal p.a. (by 1899) by establishing their own zamindari departments, buying up surface rights, including whole villages, forcing the small tenants to either work partly in their mines (while continuing to pay rent) or to leave their homes and land.[21] The total labour force, including women and children reached around 30,000.[22]

The coal mines had a share of about 5% of all rupee capital invested in industrial enterprises in Bengal. The weight of the coal mining sector was so negligible that a small government office "Geology and Mines" (within the Revenue Department) was sufficient to deal with this industry and supply statistical data on it from 1878/79 onwards.

2.2 *The period of rapid expansion*

The phase of the most rapid expansion the Indian coal mining industry experienced until present falls into the period characterised by Walter Hoffmann as the point of time in which "the industrialisation of British India" began, i.e. "from the 1890s until shortly after the turn of the century, so that we can talk of an 'industrial revolution' at this point of time".[23] From the early 1890s until the end of World War I the production of coal in India increased tenfold, the value of coal raised twentyfold, the capital invested in Indian coal mining twelvefold and the number of labourers employed fivefold.

The industrialisation in which—according to Hoffmann—the cotton and the jute industry played the most important role until, after 1911, the iron industry gained in importance, was, however, not supported by a general growth of the industrial sector. On the contrary, a number of scholars characterise the period from 1875 until 1920 as a period of de-industrialisation, ruralisation, peasantisation or one of arrested development.[24] This contradictory development is reflected in the growth of the coal mining industry.

The main features of the development of the production, consumption, import and export of coal in India can best be characterised by contrasting

[21]On March 25th 1892 the subdivisional officer in Raniganj reported to the collector cf Burdwan district on "the practice amongst colliery owners of granting land to miners". Cf. Bengal Government. Revenue Proceedings. Miscellaneous Branch. April 1892, B22, (West Bengal State Archive).

[22]This figure gives an approximate picture of the number of work days in Indian coal mining at that time and is but a rough estimate.

[23]Hoffmann, Walther: *Stadien und Typen der Industrialisierung. Ein Beitrag zur quantitativen Analyse historischer Wirtschaftsprozesse*. Jena 1931, p. 87.

[24]Cf. Thorner, Daniel: "De-Industrialisation" in India 1881-1931, in: *Premiere Conference d'Histoire Economique*, p. 217 n.

them with the development in another country, belonging to the same hemisphere, and, therefore, connected with that of Indian coal mining through sharing a number of export markets: Japan. Indeed, the Japanese development provides a guideline for comparisons. In the statistical notes on the production of coal in India, the development was always measured against the Japanese. By the end of World War I, India's production of coal, 22 million tons p.a. equalled that of China, but clearly lagged behind that of Japan (roughly 30 million tons). In 1890 the base had been fairly equal: Japanese production amounted to 2.6 million, Indian production to 2.2 million tons. The Japanese rate of growth up to 1920 averaged a little more than 8% p.a., the Indian a little less than 8%.[25] In spite of the similarity the patterns behind this development were quite different. While Japan underwent an industrial revolution leading to sustained growth, the Indian development was arrested and led nowhere near a selfsustaining industrialisation.

The difference becomes apparent when we look at coal consumption in both countries, Indian coal consumption increased at an average rate of 7%, the Japanese growth rate almost reached 10%. In 1890 Japan had raised much more coal than was consumed in the country and in the course of the three decades up to 1920 the difference between the quantity consumed and that raised became less, but consumption never surpassed production. India, on the other hand, produced more coal than it consumed until after the turn of the century and from then on consumed more than it produced.

Looking at the long term trends of development, the Japanese based on a rapidly increasing home demand, leading to decreasing (though still relatively large) shares of exports and an increasing share of imports. In India, imports were increasingly substituted by Indian coal and the exports (though relatively small in absolute figures) grew rapidly, without ever reaching a level to stimulate growth to any significant extent. The growth potential of the coal mining industry in the two countries was, therefore, very different indeed.

Fluctuations around the long term trend of development show similarities. The coal boom in India, reaching its peak in 1908, the short depression from 1909 until 1911, the shortlived boom of 1913, the relative depression throughout the war years, the boom of 1919 and the depression setting in 1920, have their parallels in Japanese production, though with time lags. These fluctuations were much more violent in India and showed a tendency to increase. Especially the depression in coal production in 1920 was much stronger in India than in Japan. So, coal production in India was characterised by strong fluctuations. In view of the insignificant

[25] These growth rates have been calculated from the notes on the production and consumption of coal in India, published from 1907 onwards. These periodically published notes provide the quantitative basis for this chapter.

GRAPH 1

COAL PRODUCTION IN INDIA AND JAPAN 1890–1920

(Semi-logarith. graph)

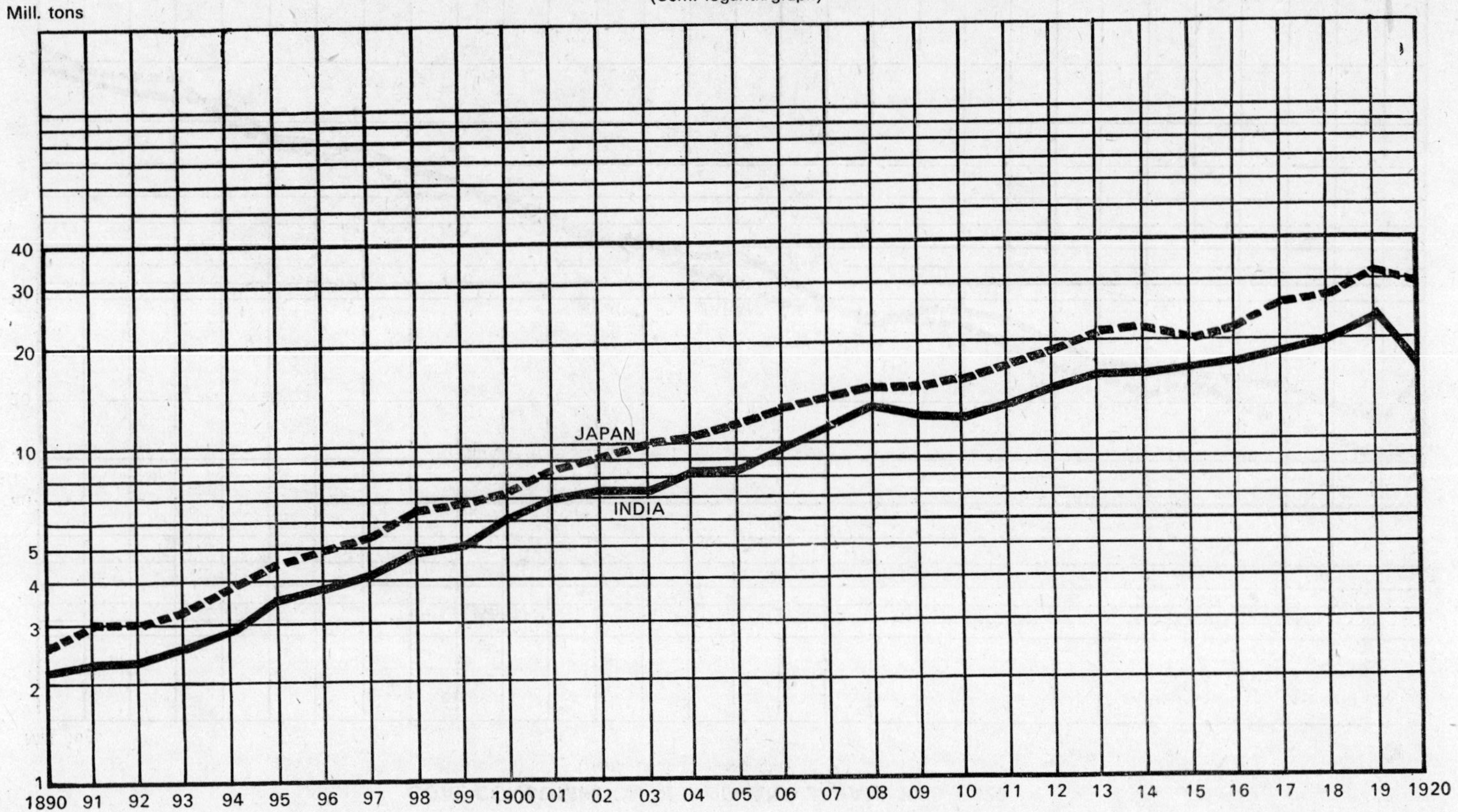

GRAPH 2

COAL CONSUMPTION IN INDIA AND JAPAN 1890–1920

(Semi-Logarithmic. graph)

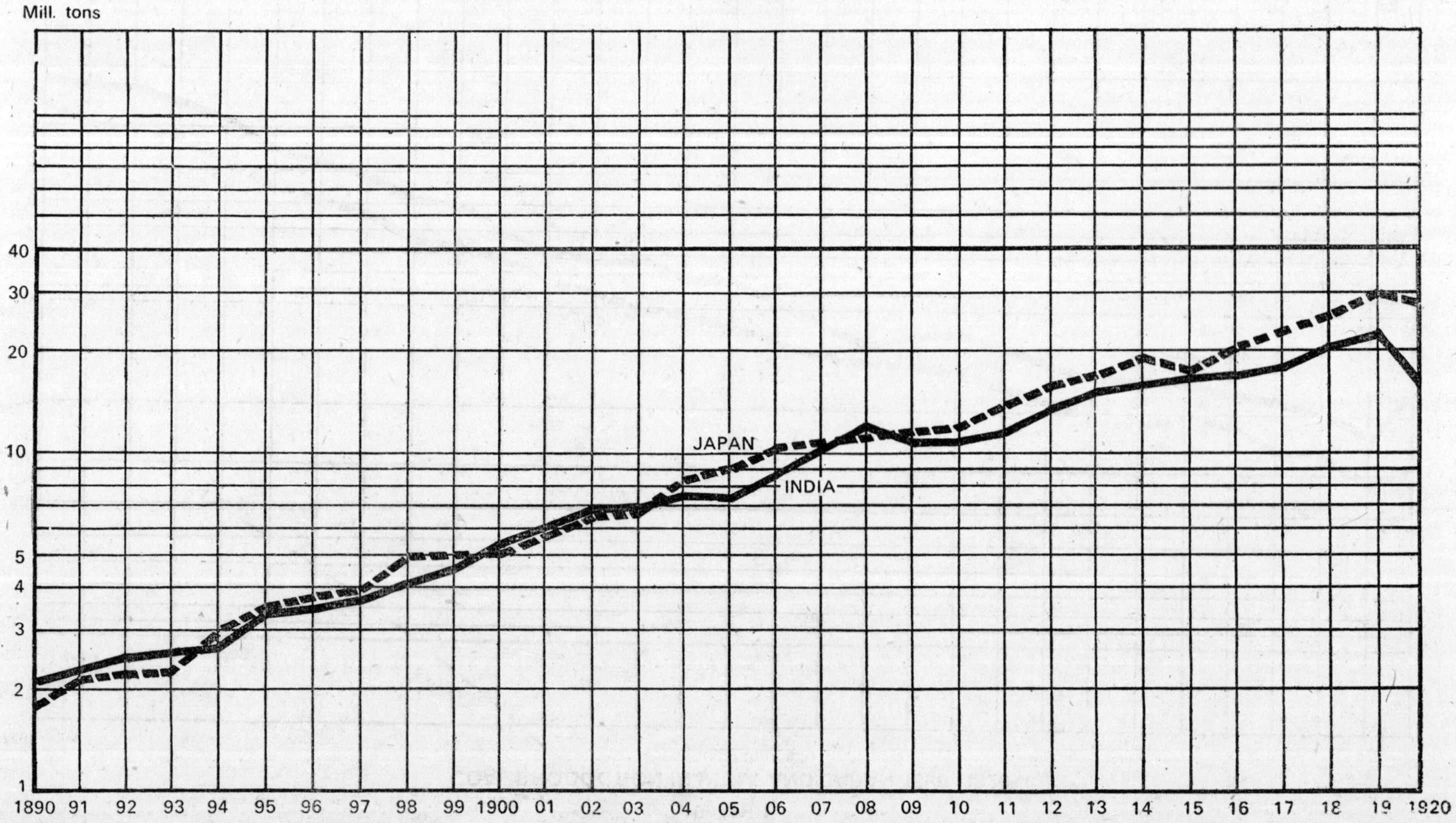

GRAPH 3

COAL EXPORTS IN INDIA AND JAPAN 1890–1920

(Logarithmic graph)

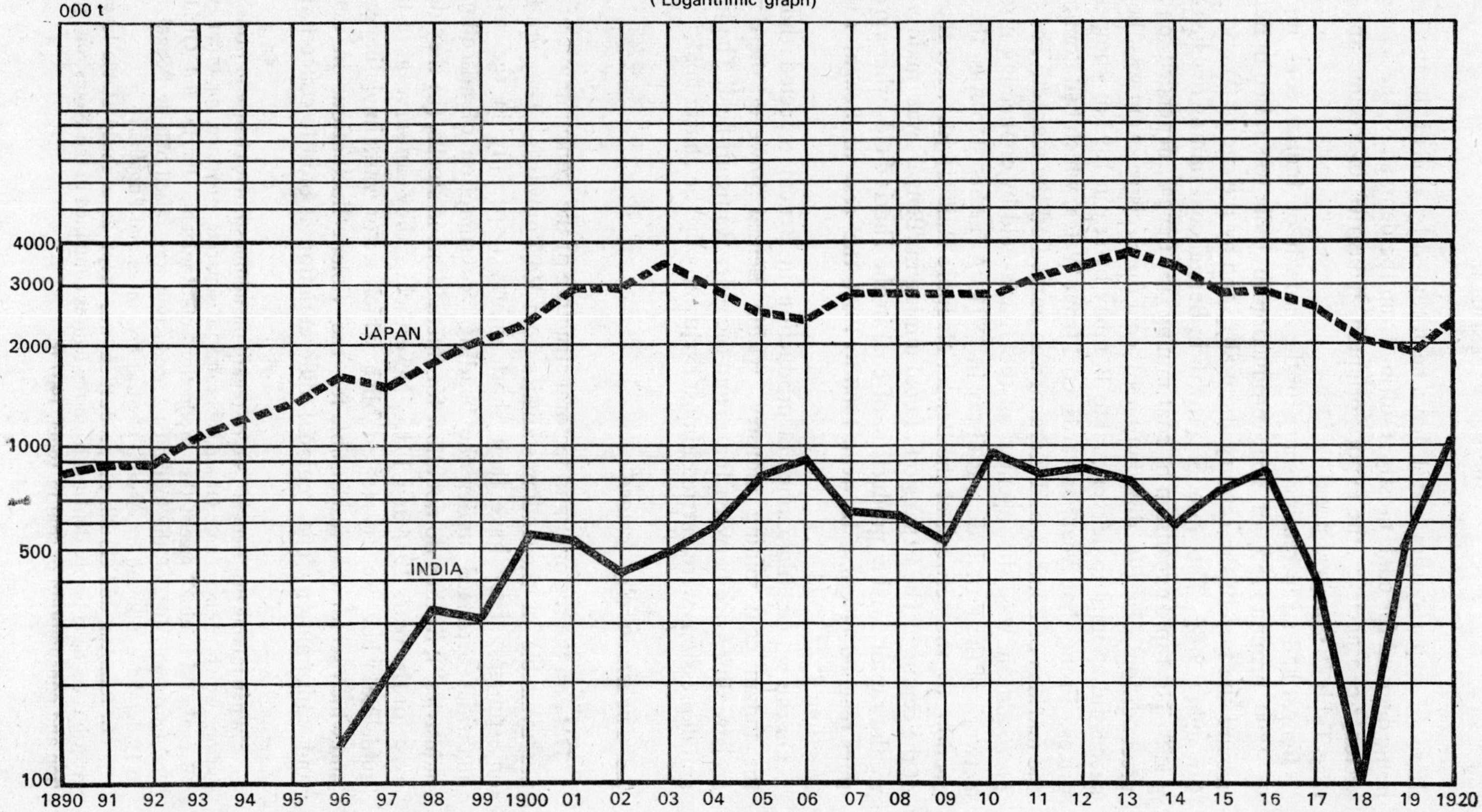

share of exports, they must have been caused by violent fluctuations in home demand for coal, mostly resulting from the demand of the secondary sector, i.e. especially the export oriented industries of cotton and jute (See Table 7.1 on next page).

Bengal had an increasing share in the growth of Indian coal mining. In 1890 around 75% of coal production in India was raised in Bengal, around the turn of the century about 80% and by the end of World War I roughly 90%. The rate of growth in Bengal coal mining reached 90% p.a.[26] The rapid expansion of coal mining in Bengal depended on two preconditions. In the distribution of coal the decisive event was the first successful attempt to deliver coal to Bombay by sea in 1893 at competitive prices. From that year onwards coal from Wales, which had dominated the markets of Bombay and Karachi was pushed off the market. The other precondition was the opening of the Jharia-field by connecting it to the network of the East Indian Railway in 1894. A further step was the connection to the network of the Bengal Nagpore Railway just after the turn of the century. The expansion of coal mining in Bengal was mainly due to the rapidly rising production of coal in the Jharia-field, where growth rates reached 11% p.a. between 1900 to 1920, the rest of Bengal a little less than 4%.[27]

On the whole, fluctuations of production in Bengal reflected those of the Indian coal mining industry. Within Bengal, however, there was a decisive difference between Jharia and the older Raniganj-field. The booms and depressions were carried almost exclusively by the Jharia-field.

2.3 *The structure of demand*

Data on the regional and sectoral structure of the consumption of coal are available in the Indian statistics from 1907 onwards. They are based on estimates. They have their shortcomings (especially in the residual quantities arranged under the heading "consumption of miscellaneous industries"); (i) they do not allow for fluctuations in stocks, (ii) they do not exist for the years 1914 and 1915. Nevertheless, they can serve as a useful guideline when we try to establish which sector was stimulating growth and which sector was responsible for the violent fluctuations in demand and consumption. The regional differentiation is of little concern in our

[26]When talking of "Bengal" we disregard the administrative change in British India during which "Bihar and Orissa" became a separate province, in other words: "Bengal coal mining" means the production of Raniganj, Jharia and Giridih. In 1907/08 roughly 10% of the Indian coal production was raised outside Bengal, mainly in Hyderabad, followed by Assam, Central Provinces and Punjab.

[27]In Raniganj and Giridih the production was kept at a fairly stable level. The growth rates have been calculated from the figures contained in the Notes on production and consumption of coal in India, 1907 ff.

context. The main bulk of coal consumption originated from Bengal, followed by Bombay and Madras (See graph 4 on next page).

TABLE 7.1
COAL PRODUCTION IN INDIA AND BENGAL, 1890-1920
(*in 1000 tons*)

Year	*India (Provinces of British-India & Princely States)*	*Bengal*	*Bihar & Orissa*	*Jharia*
(1)	(2)	(3)	(4))	(5)
1890	2.168,5	1.626,2		
1891	2.328.6	1.747,1		
1892	2.537,7	1.920,1		
1893	2.562,0	1.902,9		
1894	2.823,9	2.035,9		
1895	3.540,0	2.716,2		
1896	3.863,7	3.037,9		
1897	4.066,3	3.142,5		170,3
1898	4.608,2	3.622,1		497,7
1899	5.093,3	4.035,3		605,3
1900	6.118,7	4.978,5		650,1
1901	6.635,7	5.487,6		1.100,2[2]
1902	7.424,4	6.259,2		2.420,8
1903	7.438,4	6.361,2		2.493,7
1904	8.216,7	7.063,7		2,889.5[3]
1905	8.417,7	7.234,1		3.070,6
1906	9.783,3	8.617,8	5.325,3[1]	4.076,6
1907	11.147,4	9.993,3	6.487,6	5.179,2
1908	12.769,6	11.559,9	7.992,4	6.458,6
1909	11.870,1	10.660,8	7.134,6	5.832,7
1910	12.047,4	10.778,5	7.041,2	5.794,5
1911	12.715,5	11.468,9	7.610,3	6.373,7
1912	14.706,3	13.432,5	9.126,4	7.653,3
1913	16.208,0	14.877,6	10.227,6	8.606,2
1914	16.464,3	15.085,7	10.661,1	9.146,7
1915	17.103,9	15.693,7	10.718,2	9.140,8
1916	17.524,3	15.760,1	10.767,7	8.950,3
1917	18.212,9	16.564,0	11.932,4	9.675,1
1918	20.722,5	18.982,3	13.680,0	10.762,8
1919	22.628,0	20.897,4	15.119,8	11.969,0
1920	17.962,2	16.183,2	11.975,7	9.159,2

[1]Production figures which are listed separately for Bihar and Orissa from 1905 onwards are included in the figures for Bengal to ensure continuity. All data are taken from: Note on the Production and Consumption of Coal in India up to the year 1906 (ff.). Commercial Intelligence Department, Govt. of India, Calcutta 1907 (ff.) (appearing annually from 1906 onwards). From 1913 onwards: Report on the. . . 1913 (ff.), Department of Statistics. Govt. of India, Calcutta 1914 (ff.).

[2]Data up to 1900 are incomplete, as stated in the statistics.

[3]Data for 1901 till 1904 have been corrected later on. See Dept. of Statistics, Govt. of India: Report on the Production and Consumption of Coal in India in the Calendar Year 1913 (ff.), Calcutta 1914 (ff.).

Much more important is the sectoral differentiation. On the whole, the consumption of the secondary sector, including the consumption of the coal mines themselves, was not more important than that of the tertiary sector. But, and this difference is important for our detailed analysis later, the consumption of the railways and steamer lines oscillated around the long term trend of production. The cyclical fluctuations around the long term trend were the result of strong fluctuations in the consumption of the modern industrial sector. Both booms, that of 1908 and that of 1913, were the result of rapidly increasing demand for coal in the industrial sector, so were the depressions of 1911 and 1916.

Which branches of the industrial sector were then the decisive ones? The coal mines themselves consumed roughly 10% of the coal they raised for their own needs of steampower and later electricity. Through electrification their share rose to around 12% of the total production by the end of World War I (See Table 7.2).

TABLE 7.2
COAL CONSUMPTION OF DIFFERENT INDUSTRIES OF THE SECONDARY SECTOR IN INDIA, 1906-1920

Year	*Jute industry*		*Cotton industry*		*Iron industry*		*Construction*		*Coal mining*		*Miscell.*	
	1000 tons	%	*1000 tons*	%	*1000 tons*	%	*1000 tons*	%	*1000 tons*	%	*1000 tons*	%
(1)	(2)	(3)	(4)	(5)	(6)	(7)	(8)	(9)	(10)	(11)	(12)	(13)
1906	500	6	610	7	—	—	—	—	682	8	2.695	33
1907	667	6	794	7	131	1	—	—	1.100	10	2.824	26
1908	635	5	843	7	528	4	—	—	1.300	10	3.879	31
1909	703	6	862	7	528	4	—	—	1.131	10	3.110	26
1910	754	7	870	8	374	3	843	7	1.204	11	1.768	16
1911	623	5	872	7	407	3	890	7	1.272	10	2.077	17
1912	724	5	1.005	7	625	4	1.029	7	1.471	10	2.890	20
1913	769	5	1.171	7	760	5	1.135	7	1.621	10	3.643	23
1914									1.898	12		
1915									1.921	12		
1916	912	6	905	6	843	5	440	3	2.071	13	2.993	18
1917	1.026	6	927	5	807	5	358	2	2.186	12	3.226	18
1918	1.045	5	968	5	1.307	6	477	2	2.487	12	5.126	25
1919	891	4	1.033	5	1.262	6	520	2	2.715	12	6.408	29
1920	1.085	7	1.069	6	1.416	8	413	2	2.335	14	942	6

Source: Note on the production of Coal in India. . ., loc. cit., Calcutta 1907 ff.

Until 1917 the jute and the cotton industries were the largest industrial consumers of Indian coal. The jute mills and cotton mills in the table include only spinning and weaving mills and exclude the cotton and jute presses, which consumed a considerable amount of coal shown under con-

GRAPH 4

DEVIATIONS FROM TREND OF CONSUMPTION, SECONDARY & TERTIARY SECTOR IN INDIA 1906–1920

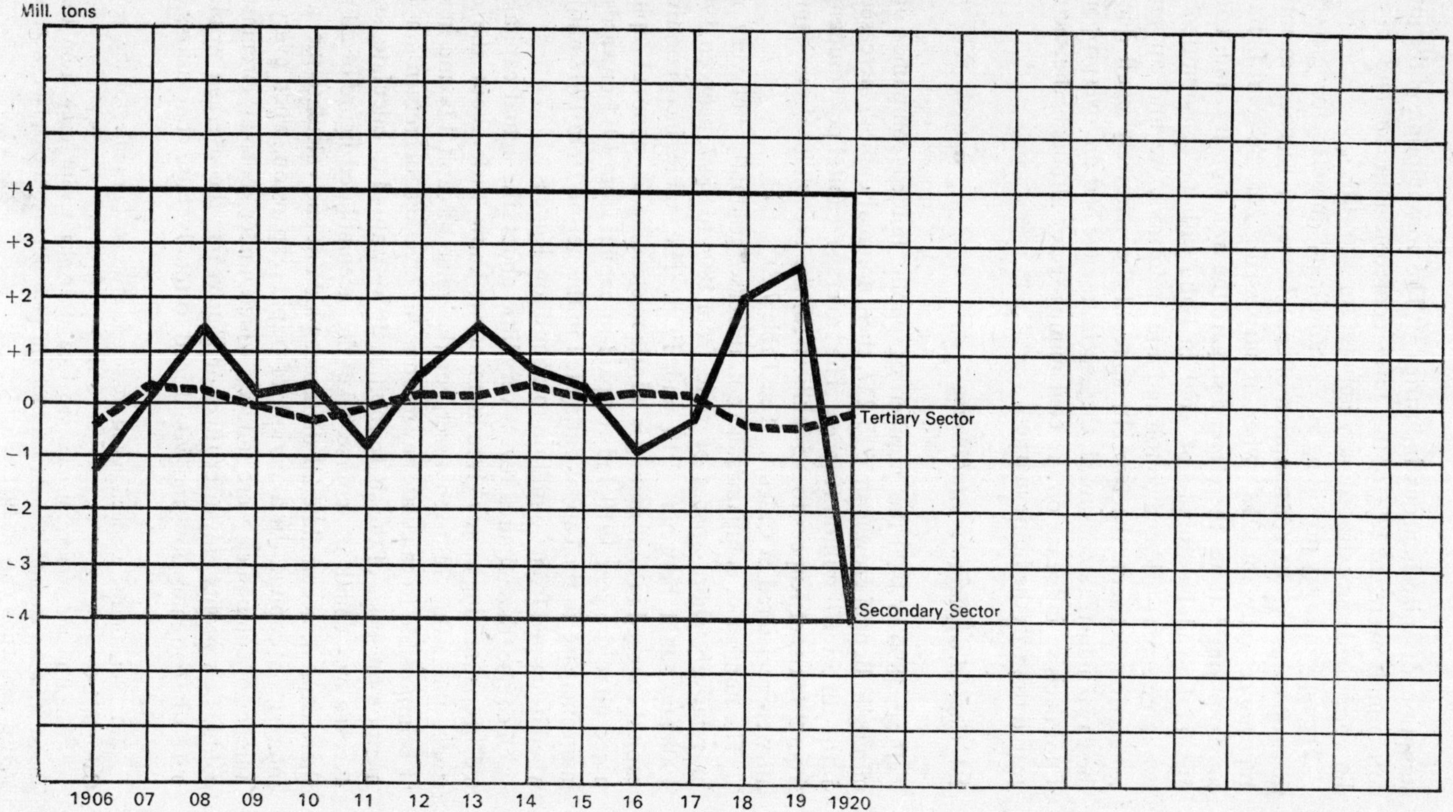

sumption of miscellaneous industries.[28] The consumption of the jute and cotton mills averaged 12% of the total coal consumption and obviously did not provide an extraordinary impulse for growth.

The only industry providing an extraordinary impulse for growth was the iron industry. Its consumption of coal, however, reached quantitative importance only in 1918, when it surpassed 5% of the total Indian coal consumption.[29] If we, therefore, disregard the iron industry and correlate the total available supply of coal in India with the consumption of the two large export-oriented cotton and jute industries, we find an amazing connection between the two. In spite of the relatively small quantities these two industries consumed, they seem to have had an overproportional if not decisive influence on the coal mining sector although the sector was geared towards the home market.

2.4 *Capital employed in coal mining*

From 1890 to 1920 the number of public limited companies in coal mining in Bengal grew from 6 to 227, their nominal capital increased from 5 to almost 87 million Rupees. The average capital per company sank from 9 lakhs of Rupees to 3.5 lakhs. With the exception of two companies all were registered in Calcutta (See Table 7.3).

The number of companies did not increase gradually but very rapidly in relatively short spaces of time, from 1894 to 1897, between 1907 and 1908 and again between 1918 and 1920. The table incorporating the capital engaged in coal mining is a little misleading, as nominal capital was noted up to the year 1910 and the paid-in capital from 1911 onwards. The main increase in the capital outlay fell in the period from 1907 until 1909, when it jumped from 26 million to 55 million Rupees.

Of course, these figures reveal very little about the actual capital invested in coal mining at any given moment, we know little about the capital stock, the value of fixed assets or the value of capital bound in stocks. It is impossible to calculate any meaningful average of return on investment on the basis of the share capital. If we wanted to calculate such an average we could of course take the accumulated figure of all paid in share capital, accumulate all dividends paid out and calculate the average interest on capital. In a period of rapid expansion, when every year a number of companies were added, which under normal circumstances, could show profits only after three or four years the average would turn out to be extremely low and would lead to totally false conclusions.

[28]Cf. Note on the production and consumption of coal in India for the year 1907, Calcutta 1908, p. 4.

[29]The iron and steel companies provided their plants with coal from their own mines. Tata for example bought Jamadoba in 1906 although their plant in Jamshedpur started production only in 1911.

TABLE 7.3
SHARE CAPITAL IN COAL-MINING OF BENGAL, 1890-1921

Year	Number of Joint-Stock Companies	Nominal Share Capital in 1000 Rs.	Paid-in Share Capital in 1000 Rs.	Debentures in 1000 Rs.	Actual total Capital in 1000 Rs.
(1)	(2)	(3)	(4)	(5)	(6)
1890[1]	6	5.270	5.195	300	5.495
1891	10	6.355	5.229	279	5.508
1892	9	7.345	5.914	422	6.336
1893	12	7.795	6.111	925	7.036
1894	14	8.645	7.579[2]	950	8.529
1895	18	9.775	7.810	950	8.760
1896	25	10.954	9.301	930	10.231
1897	31	13.634	9.886	930	10.816
1898			10.758[3]	3.256	12.052
1899	32	15.111	11.936	1.294	13.230
1900			10.847[3]	1.275	12.067
1901			11.247	1.220	13.144
1902			14.157	1.897	14.286
1903			16.410	129	17.436
1904			16.423	1.026	17.449
1905			17.802	646	18.447
1906	48	28.704	24.028	1.596	25.624
1907	64	40.079	25.924	4.644	30.567
1908	112	68.139			
1909	121	73.933	54.888	5.845	60.733
1910	119	74.199	60.950	2.474	63.424
1911	120	62.309[4]	60.556	3.162	63.718
1912					
1913	124	56.850	56.154	1.467	57.622
1914	129	57.595	57.072	1.347	58.419
1915					
1916					
1917	151	63.993	63.055	1.796	64.852
1918	165	69.724	69.134	2.283	71.417
1919	183	73.778	73.062	1.732	74.795
1920	227	76.198	75.519	1.783	77.302
1921	247	86.838	85.774	2.108	87.881

[1]Position as per end of March of the respective year.

[2]The capital of the Equitable Coal Company which was newly founded in 1895 has been considered as paid-in capital.

[3]Data for 1898 and for the period from 1900 to 1905 were taken from the journal "Capital", Calcutta 1898 ff. Position as per 30.6. of the respective years; data for 1890-97 and 1899 were taken from the Reports on Joint Stock Companies, published in the Proceedings of the Revenue Department of Bengal, Miscellaneous Branch; from 1906 onwards they were taken from: Note on the production and consumption of coal in India 1906 (ff.), Calcutta 1907 (ff.). From 1906 onwards the capital of the joint-stock companies established in London is included.

[4]Up to 1910 the "Nominal Capital" comprises the authorized, from 1911 onwards the signed share capital.

So the only way is to select. Again, the process of selection would have to be carefully laid out: It seems doubtful that we can follow A.B. Ghosh's method of taking at random the dividends of 7 or 8 extremely successful coal companies and arguing about profits in Indian coal mining without examining the background of the data.[30] One point should be stressed in any case. The dividends reveal very little about the profits. It is wrong to estimate the rate of profit by looking at the dividends. The facts that dividends were kept stable in many companies (the Bengal Coal Company held their dividends at a level of 12% p.a. for decades) shows that large reserves were accumulated to level out differences in times of less prosperity.

2.5 *Labour employed in coal mining*

In 1890 around 25,000 labourers were working in the coal mines of Bengal, or to be more precise, the number of working days was equivalent to that of a permanent labour force of 25,000. Up to the rapid expansion of coal production in the 1890s, a semi-feudal system of recruitment and control of labour had prevailed. Scattered data[31] suggest that roughly a quarter of the miners employed in Raniganj held leases issued by the zamindari departments of the coal companies with the only condition of work in the mines (this seems to have been the permanent work force), others held sub-leases on the condition of working a number of days in the coal mines and paying a rent at the same time. The rest of the labour force was recruited from adjoining villages and the "supply" of labour depended very much on the harvest and on weather conditions. The worse the harvest, the more landless labourers and small farmers were seeking employment in the coal mines during the dry season. In the Giridih coal-field the total labour force was recruited from an area not larger than 20 miles around the mines. A report written in 1921 states that not a single worker out of the total number of 8,000 miners migrated to the mines. A fairly detailed analysis of the work force in Giridih from 1891, resulting from an enquiry into legislation for the women and children working in coal mines, gives an indication of the average structure of the labour force. Nearly 5,000 workers were employed in the railway collieries, earning a living for themselves and in addition for 3,500 dependents. Half of them worked underground, the other half on the surface. (See Table 7.4 on next page).

If we follow the Labour Commission of 1896 in their estimate that the monthly wage of a miner in Sanctoria (Raniganj) averaged 5 1/2 to 6 Rupees

[30]A.B. Ghosh: 'Coal Industry in the Pre-Plan Period: Trend of Profits and Investment', in *Economic and Political Weekly*, Bombay, 25.10.1969.

[31]Hints on these questions are contained in the *Report of the Labour Enquiry Commission*, Calcutta 1896, and Blanchard Foley's *Report on Labour in Bengal*, Calcutta 1906.

throughout the period from 1880 to 1894, while the prices for rice were rising by 53% (between 1880 and 1884 one Rupee bought 23 seers of rice on an average, between 1890 and 1894 a little less than 15 seers) and that the monthly wage of an "able-bodied" agricultural labourer in Burdwan sank from 9 Rupees in 1880 to 7 Rupees in 1894,[32] two things become apparent; (i) real wages were sinking, (ii) coal mining did not provide material incentives for labourers, therefore, obviously, other means had to be employed to bring workers into the mines. If a miner was not induced to drinking, the Commission of 1896 estimated, he could earn his week's living by 4 to 5 working days' wages.[33]

TABLE 7.4
WORKERS UNDERGROUND

	Men %	*Women* %	*Children* %	*Total* %
Cutter	44	2	1	47
Loader	5	32.5	10	47
Others	5	0,5	—	6
Total	54	35	11	100

Concerning the 2516 workers on the surface there was a similar distribution

Men	Women	Children
57%	35%	9%

From 1890 to 1919 the number of workers in the coal mines of Bengal increased from 25,000 to 175,000. The number of children employed in coal mines remained almost constant, while the number of women employed increased more rapidly than the number of men. Productivity rose steadily from 1890 to around 1906, from 66 tons per man per year to 110 tons but remained constant from then onwards, indicating the lack of technological change and investment in the second half of our period.[34]

We do not know how the structure of the work force changed in the course of this expansion. An industrial census, taken in the year 1921 reveals where the development led to. In 1921 the number of people employed in coal mines in Bengal and Bihar & Orissa was given as 150,030. Of the 82,619 employed in Manbhum 347 or 0.5% were employed in managerial positions, 2% technical personnel and supervisors, 2% administrative and clerical staff, 39.5% were classed as skilled and 56% as unskilled labourers. The distinction was obviously between work underground and on the surface.[35]

[32] *Report of the Labour Enquiry Commission*, Bengal Secretariat Press, 1896, p. 7.

[33] Ibid., p. 20.

[34] An analysis of actual investments within the group of coal mines managed by Andrew Yule confirmed this result.

[35] *Census of India*, Vol. I, India, part 1 (1921), p. 248.

In 1917 it was estimated that 15% of the miners in Jharia were "settled," most of them Santhals, 75% worked in gangs migrating to Jharia for a number of weeks or months, and that 10% came from surrounding villages "daily or whenever it suited them".[36] In Raniganj conditions were quite different: about 50% were classed as "settled", obviously workers in the old established pioneer companies with large zamindari-departments; of the other half 90% came to the coalfields in gangs like in Jharia and 10% came irregularly from villages in the not too distant neighbourhood.[37] The coalfields of Rajhara and Giridih, both in fairly isolated situations remained in an exceptional position. In Giridih the advantages of certain rudimentary features of social security were combined with a stricter control over the labour force than could be found elsewhere.[38]

It is difficult to give any clear indication of the development of nominal and real wages. From scattered data it seems that real wages of miners showed a slight decrease from 1894 to 1906, and increased slightly after 1908. From 1917 onwards evidence strongly suggests a sharp fall in the living standards of miners—nominal wages fell sharply in 1919 while prices for the cost of living doubled between 1914 and 1922.[39] In any case, the cost of labour did not determine the price of coal and the ratio wages versus companies' returns was decreasing. So there was nothing to suggest labour saving innovations.

2.6 *Technological stagnation*

None of the factors we have been analysing so far suggested that any significant technological change took place in Indian coal mining between 1890 and 1920 (with the exception of a small number of mines with deep shafts). This conclusion is verified if we look at the size and technology of a "typical mine" in 1919. In this year 587 mines were being worked in Raniganj and Jharia, raising 18.8 million tons. Ca. 50% of these mines had a monthly production of less than 1,000 tons, 35% between 1,000 and 5,000 tons and 15% more than 5,000 tons. This last group accounted for more than 60% of production, the middle group 35% and the first one 5%. The mines worked by public limited companies, accounting for 68% of coal production in this year, belonged to the number of 86 mines with an average yearly production of 1,35,000 tons per year.

43% of all mines were being worked without any mechanical equipment, 55% were equipped with steam power, only 2% were working with electrical energy. Mines equipped with steam power accounted for 85% of pro-

[36] Cf. *Census of India* Vol. I, India, Part 1 (1921), p. 249.

[37] Cf. ibid., p. 250.

[38] Cf. Blanchard Foley, *Report on Labour in Bengal*, 1906, p. 33.

[39] Cf. 'Index Numbers showing the Rise and Fall in the Cost of Living in Bihar and Orissa', in: *Bulletin of the Department of Industries*, Bihar and Orissa, 1922/23 ff.

duction, the electrified ones for 7%. Only 4 mines, 2 in Raniganj and 2 in Jharia were equipped with cutting machines, all other mines had very little technical equipment indeed.[40] A "typical" mine of the "modern" type seems to have consisted of four inclines and two shafts, working one or two seams. On any of the inclines roughly 2,000 tons of coal were raised through the shafts (average depth 400 feet) around 4,000 tons of coal. Per incline one boiler was needed and a winch being able to pull about 7 tubs of coal containing 2/3 of a ton each. The lifts in the shafts were normally driven by one or two Lancashire boilers. On an average, this type of equipment seems to have been incredibly wasteful. Rentability calculations revealed that theoretically 51 tons of rubble would have sufficed to raise the quantity of coal actually produced and would have provided enough energy for the water pumps, while in effect 1,320 tons were used, equaling 8% of total coal production.[41]

The management of the mines was practically reduced to the small number of Europeans and skilled workers needed to do maintenance work on this equipment and to decide which seams to work on. The labour management was normally delegated to the contractors and was thus no part of the management of the mine itself. In the course of expansion the tendency seems not to have been to incorporate labour management into the mines, but on the contrary, companies like the East Indian Coal Company, starting off with direct labour management changed over to the system of raising contractors after a few years.[42]

From the evidence we can conclude that there was no concentration of production in larger units, there was no significant tendency towards mechanisation, there was no cost pressure leading to rationalisation, but rather a hesitation to invest any capital at all even at the cost of wasteful production. A permanent "modern" labour force did not emerge, productivity increased only during the initial phase, up to 1906, possibly through investments in raising and pumping equipment. Modernisation on the whole did not take place.

3. *The Structure of the Managing Agency System*

In the course of the four decades from 1880 until the end of World War I the managing agents of Bengal changed their role. In the beginning

[40]Cf. *Report of the Coalfields Committee*, Calcutta 1920. "Relevant Statistical Information" (1919), on p. 200.

[41]Cf. Marshall, A.G. and W.A. Lee: 'Central Power Stations for the Coal-Fields', in: *Transactions of the Mining, Geological and Metallurgical Institute of India*, Vol. III, No. 12, Calcutta 1908, pp. 99-103; and White, J.: 'Waste in Bengal Coal Mining', in: ibid., vol. II, No. 1, Calcutta 1907.

[42]Cf. 'Company reports of the East Indian Coal Company', in: *Capital*, Calcutta, 1898 ff.

of industrialisation in Bengal, in the 1850s to 1870s, the normal pattern of establishing an industrial enterprise was for the investors (including some people who knew the business) to form an association. Since the introduction of the Company Law in Bengal this association would typically be a Limited Liability Company. They appointed the directors and it was the task of the directors to find the manager(s) or to appoint secretaries to run the business. The control of the company, in any case, was in the hand of shareholders, and the position of a director necessitated a specified percentage of the share capital to ensure a certain identity of his interests with the interests of the majority of shareholders.

In the 1870s and 1880s it became the normal feature to appoint a managing agent for the company, an agent fit to handle the day-to-day business of the company. The managing agent did not have to hold shares in the company, he had the role of a hired manager. Only that the contracts concluded between the companies and the managing agents were much more binding to the companies than contracts with employees. There are three levels at which we have to outline the function of the managing agent. The first and the most important is the role of the managing agent vis-a-vis a single company; the second is the role within a group of companies; and the third (in our context least decisive one) the role within the network of international capital relations.

The main aspect of the managing agent's role within any limited company was the discrepancy between the interests of the shareholders and their directors and the interests of the managing agent. This contradiction became apparent on two levels, firstly the terms of contracts and secondly the position of the company as the employing party vis-a-vis the managing agent. The typical form of remuneration of a managing agent in the early period was a commission on the sales of the company, irrespective of the company's profits. Thus, the managing agent's aim could easily be to maximise sales, while neglecting profits and dividends. The range of possible manipulations by the executive within this framework is quite obvious and it was widely used.

It must have been the reluctance and the active protest of the investors against this type of agreement that brought the change to the practice which becomes usual in the 1890s and after the turn of the century. Newly established companies, especially in coal mining, usually paid the managing agents a percentage of their profits. But even where the commission was paid on the basis of "net profits" the interests were not identical. For once, net profits were calculated before depreciation on investment and funds put to reserves. So, even under this type of arrangement, earnings of the managing agents and the investors (via dividends) could and did differ widely. Secondly, there is no evidence which would lead us to assume that there was "a natural tendency" of managing agents to level out the profits of all jute mills, or coal companies under their control.[43] We only have to look at data, which are readily available to falsify Lokanathan's conclusion:

TABLE 7.5
DIVIDENDS FROM FOUR JUTE FIRMS OF THE MANAGING AGENCY ANDREW YULE & CO., 1881-1910

Period	Year founded	*Budge Budge* (*1873*)	*Central* (*1890*)	*Delta* (*1898*)	*National* (*1895*)
1881-1885		10.6%			
1886-1890		7.7%	4%[1]		
1891-1895		16.6%	5.4%		
1896-1900		8.4%	6.2%		8.3%[3]
1901-1905		12.5%	10.8%	10%[2]	8%
1906-1910		20.6%	11.5%	4.8%	8.8%

[1]Only in 1890; [2]Only in 1904 and 1905; [3]Only in 1898-1900.
Source: *Investor's India Yearbook*, loc. cit., Calcutta 1911, p. 72 ff.

Dividends, as shown in the above table, showed a tendency to the opposite. Quite obviously there were only a few companies showing high profits during boom periods. In our graph the profits of the Budge Budge Jute Mills almost doubled from 1901-05 to 1906-10, those of the other companies remained stable (Central and National Jute Mills) or rank even (Delta Jute Mill). The main problem, however, was not the type of commission the managing agent received but the independence or dependence of a company vis-a-vis the agent. Most companies in the 1890s and after the turn of the century were founded by the agency houses of Calcutta themselves. They started off with a contract binding them to the agency. The success of the company always depended on its position within the group of companies under the agent's control. The chances for success of any newly established company in Bengal in the 1890s were judged by the composition of the companies under the control of the same managing agent. When the Bengal Nagpore Coal Company thought of seeking a new managing agent the chances of Jardine Skinner & Co. were described as follows: "With two jute mills and a paper mill they should be able to influence business"[44] or Finley, Muir & Co. "with the business they can influence, being agents for two jute mills, a line of steamers, a press and a paper mill, the company should do well."[45]

A certain degree of vertical integration guaranteed a good demand for coal, a certain degree of horizontal concentration guaranteed the necessary

[43]This is Lokanathan's central thesis, cf. Lokanathan, op.cit., pp. 290-296.
[44]Cf *Capital*, Calcutta, 16.11.1892, p. 379.
[45]Cf. *Capital*, Calcutta, 14.12.1892, p. 455.

know-how and distribution facilities in the industry, a high number of companies under their control a substantial financial backing. The larger the involvement of the managing agent, the stronger was his position. Indeed, the balance between the managing agents and the individual companies shifted rapidly to the position of strength on the side of the agency houses and weakness on the side of the company's shareholders, who were often split into two groups, one in Britain, one in India. The decreasing power of shareholders was reflected in the fact that companies with limited liability launched by Andrew Yule after the turn of the century did not even have a board of directors.

A second factor contributed to this development. In an attempt to mobilise capital in Bengal on all levels, the nominal value of shares sank rapidly. In the 1850s the Bengal Coal Company had issued shares with a nominal value of Rs. 1,000; around the turn of the century the average nominal value of shares had sunk to Rs. 50 or even Rs. 10. This meant an increasing distribution of shares among larger numbers of small shareholders. The wider the distribution of shares, the smaller became the necessary minimum capital for controlling the company. The managing agency houses were thus in a position to control capital which they did not possess. Their control originated from their know-how, their business relations in Bengal and abroad, their easy access to credit, either through their own banks or *via* commercial banks to whom they could give guarantees beyond the means of an individual company. But in the long run this control could only be maintained if the managing agents acquired controlling share majorities in key companies.

The integration on a vertical and on a horizontal level expressed itself in the organisational integration of companies within departments of the managing agency house. There was a coal department, jute department, tea department etc. coordinating the activities in each industrial branch. It is quite obvious that companies of the same industry did not compete with each other, because distribution and sales policy was centrally controlled by the respective departments. On the vertical line, the picture was more complicated. There were some industries in Bengal, which had no interrelations, like tea and jute. The coal mining industry was, however, in a different position. Coal was needed in almost all of the typical industries of Bengal, so the demand for coal and the price of coal etc. had to be coordinated. The only principal guiding line for the operations of the whole group could be the maximization of profits of the managing agent, which could, for obvious reasons, run quite contrary to the interests of an individual company. *Via* the price policy, profits could easily be shifted from one company to the next, from one industry to the other.

The core of the managing agency was always the finance department which could have different organisational patterns. The usual feature in the large agency houses was a bank, in the case of Andrew Yule & Co., formerly the Bank of Calcutta, later the Mercantile Bank of India Ltd. Finally, the managing agencies were the coordinating bodies of the international flow of capital, between Britain and India on the one hand, and between India and other British colonies and export markets on the other. As the managing agents were private companies with changing business partners, which did not publish their balance sheets, it is impossible to reconstruct their role in the international framework without access to the carefully guarded legacies of men like David Yule, or Ernest Cable *et. al.* The first business history of Bird & Co, published in 1929, gives a vivid impression of the way of thinking, the aggressiveness and the flexible coordination of various interests in each of their business deals, which characterised the most successful heads of the large managing agencies.

All large managing agencies of Bengal established agencies in other parts of India and South Asia, as well as in London. Many of them were connected to British business firms through various types of agreements, as importers for British machinery, as sellers of Indian material etc. A number of the larger managing agency houses started to invest capital in England, which had been accumulated in India. It is difficult to reconstruct the growing involvement of these houses in international finance capital. For Andrew Yule & Co. some of the steps will be mentioned in the next section on the development of this business house.

4. *Andrew Yule & Co.*

Andrew Yule & Co. was one of the two largest European managing agency houses in Bengal, which controlled large industrial empires in the inter-war period. They surpassed all other Calcutta business houses in the 1890s and the first two decades of the 20th century. They started off as small agency houses, partly as representatives of large companies (insurance agents etc.) and ended up as large holding companies, controlling groups of companies covering all industries which were to be found in Bengal. The aggressive expansion of Andrew Yule & Co. was steered by David Yule, who led the firm, which had been established in 1865, from the position of a small managing agent, mainly looking after the affairs of one cotton mill, to the position of a holding and management company controlling roughly 120 industrial companies in Bengal.[46]

In the period covered by our analysis the main strongholds of Andrew Yule & Co. lay in the jute industry (in 1907 companies under its control comprised 10% of the industry's capacity in Bengal), inland steam shipping, cotton industry, flour and oil mills, tea gardens. One of the largest interests was coal mining. In this field, Andrew Yule & Co. came to a leading position and surpassed all other large agency houses.

[46]Cf. *The Statesman*, Calcutta, 5.7.1928: "Sir David Yule Dead".

The commanding position within Andrew Yule & Co. was held by the private partnership company itself and the financial department. From 1896 onwards Andrew Yule & Co. did not only describe themselves as exchange and commission agents but at the same time as bankers.[47]

During the same year the managing agency house established the Calcutta Bank which was later amalgamated with the Mercantile Bank of India Limited.

The Mercantile Bank was not an Indian bank. It had been established in London in 1892, with a capital of 1.5 million pound sterling as an international chartered bank. The takeover took place in 1906. With this transaction Andrew Yule & Co. stepped into the world of international finance capital. The Mercantile Bank did not only have agencies in Calcutta, Bombay, Delhi, Madras, Karachi and Rangoon, but also three agencies in Ceylon, and agencies in Singapore, Penang, Khota Baru, Kuala Lumpur, Batavia, Hongkong and Shanghai.[48]

The expansion of the managing agency house in other fields was paralleled by the growth of Andrew Yule & Co's coal department. In 1890 the Katras-Jherriah Coal Company was established and "successfully launched" in 1893. This company established the Seebpore Company in 1900, by selling a part of its property to the new shareholders. A second dependent coal company was launched in 1908—the Minto Coal Company. In both cases the parent company held the controlling majorities of share capital.[49]

In 1893 the Adjai Coal Company was launched. It remained formally independent throughout the period of our analysis. This company, similar to the Seebpore and the Minto Coal Co. did not have a board of directors, but was controlled directly and only by the Managing Agent. The Bengal Nagpur Coal Company had not been established by Andrew Yule & Co. Its management was taken over in 1896 after another agent had been unsuccessful. The Bengal Nagpur Coal Company expanded along the lines of the Katras-Jherriah Coal Company and founded two new companies, the North-West and the Industry Coal Co. by selling parts of their land to them.[50]

David Yule's most important *coup* in coal mining was the takeover (of the management) of the Bengal Coal Co., the largest coal company in India, during the boom year of 1908. When this group of companies (the Bengal Coal Co. had established two branches, the Bengal Bathdee and the Bengal Giridih Coal Co) was added, Andrew Yule & Co. controlled the largest

[47]Cf. *Thacker's India Directory* of 1896.

[48]Cf. *Investor's India Yearbook*, Calcutta 1916, p. 68.

[49]Cf. *Capital*, Calcutta, 13.9.1900, p. 389; *Investor's India Yearbook* 1916, pp. 276 and 283.

[50]Cf. *Investor's India Yearbook*, 1916, p. 244, and p. 272.

coal interests in India. Their group raised the largest quantity and owned the largest reserve lands.

In later years, a number of other companies, mostly founded by other agents, were incorporated into the group when they ran into difficulties. The Sudamdih Coal Co. and the Burro Demo Coal Co. are examples.[51] By the end of 1908 Andrew Yule & Co's coal department controlled around 13% of the total coal production of Bihar & Orissa, 1.4 million tons of a total quantity of 11.6 million tons raised. The group's capital reached 6.5 million Rupees, or 10% of the capital invested into coal mining in India. (Bird & Co's coal companies had a capital of 8.8 milli›n Rupees, Shaw Wallace's 13 coal companies 5.2 million, MacNeill & Co. 5 million.). By the returns from sales the coal department of Andrew Yule & Co. surpassed all others by far. The Bengal Coal Co. alone sold coal worth 7.2 million Rupees in 1908, more than any of the other groups.[52]

5. *The policy of Andrew Yule & Co in coal mining*

Of the companies controlled by Andrew Yule & Co. not all the books have been kept in the archives of the B.C.C. From the books available we can reconstruct the behaviour of the Katras-Jherriah Coal Company plus its two branches, the Seebpore and the Minto Coal Company, the Adjai Coal Company (1894-1918) and the Bengal Nagpore Coal Company (1896-1912).[53]

5.1 *Horizontal concentration*

A first look into the accounts reveals that not all of the companies distributed their coal themselves. The dependent companies do not show any expenditure under freight, loading charges etc. The key companies, on the other hand, show large sums of expenditure under the heading "coal purchases". Table 7.6 shows the sales of each company and the percentage

[51]Cf. *Capital*, Calcutta, 24.12.1908, p. 1301; and *Investor's India Yearbook*, 1916, p. 156.

[52]The second difficulty lies in the varying cost categories over the period. The accounts of the companies separated different types of costs only if they were of special interest. The interest varied over time. There were times when, for example, costs of recruiting labour mattered, there were others when they did not. In this case they were accumulated under another heading. This means that we cannot follow the development of all types of cost throughout the period under analysis.

[53]The archives of the Bengal Coal Company contains only one journal of the Bengal Nagpore Coal Co, ending in June 1912. The journals of the Minto Coal Co (from 1907 onwards), the Seebpore Coal Co (from 1900 onwards), the Adjai Coal Co (from 1894 onwards) and the Bengal Giridih Coal Co (from 1908 onwards) cover the period under investigation in full.

TABLE 7.6
SALES AND PURCHASES OF COAL WITHIN THE COAL DEPARTMENT OF ANDREW YULE AND COMPANY, 1894-1918

	Adjal Coal Company			*Bengal Nagpore Coal Company*			*Seebpore*
Year	*Total Turn-over in 1000 Rs.*	*Sales to the K.J. Company in 1000 Rs.*	*Share of K.J. Co in %*	*Total Turn-over in 1000 Rs.*	*Sales to the K.J. Company in 1000 Rs.*	*Share of K.J. Co in %*	*Total Turn-over in 1000 Rs.*
(1)	(2)	(3)	(4)	(5)	(6)	(7)	(8)
1894	74						
1895	112						
1896	170	149	88	534	46	9	
1897	181	187	100	697	88	13	
1898	183	139	76	523	88	17	
1899	163	170	100	400	39	10	
1900	226	192	85	458	31	7	9
1901	232	141	61	405	25	6	49
1902	195	169	87	419	31	7	80
1903	182	171	94	557	33	6	111
1904	205	202	99	254	45	18	140
1905	207	203	98	374	50	13	139
1906	236	225	95	452	43	10	174
1907	305	290	95	650	45	7	228
1908	292	266	91	1,008	40	4	278
1909	201	199	100	562	110	20	223
1910	232	245	100	755	170	23	246
1911	191	202	100	671	76	11	181
1912	212	218	100	660	68	10	337
1913	198	201	100		35		326
1914	224	202	90		120		301
1915	252	243	96		27		349
1916	294						304
1917	239						167
1918	392						358

Source: Turnover of the individual firms as listed in the day-books of the respective firm (B.C.C.-Archives).

The book of the Bengal Nagpore Coal Company only covers the period of 1896-1912.

The sales of the individual firms to the Katras-Jherriah Coal Company were compiled according to the cash books of the Katras-Jherriah Coal Company. They cover the period 1899-1916 (B.C.C.-Archives).

Coal purchase of the Katras-Jherriah Coal Company were taken from the day-books of this firm.

of sales that went to the Katras-Jherriah Coal Company. 96% of the Seebpore Company's sales went to the Katras-Jherriah, 100% of the Minto Company's and 93% of the Adjai Company's sales, although this

Table 7.6 *Contd.*

Coal Company		*Minto Coal Company*			*Katras-Jherriah Coal Co.*			
Sales to the K.J. Company in 1000 Rs.	*Share of K.J. Co in %*	*Total Turn-over in 1000 Rs.*	*Sales to the K.J. Company in 1000 Rs.*	*Share of K.J. Co in %*	*Sales to the K.J. Coal Co (total) in 1000 Rs.*	*Total Coal purchase in 1000 Rs.*	*Coal purchase by its own firm in %*	*Total Turn-over in 1000 Rs.*
(9)	(10)	(11)	(12)	(13)	(14)	(15)	(16)	(17)
					195	281	69	1,361
					275	397	69	1,824
					227	408	56	1,528
					209	328	64	1,122
6	100				229	418	55	1,834
51	100				217	359	60	1,344
83	100				283	343	83	1,313
118	100				322	402	80	1,580
128	91				375	620	60	2,422
140	100				393	625	63	2,754
174	100				442	658	67	2,749
199	87		1		535	932	57	3,204
246	88	36	45	100	597	1,165	51	3,795
207	93	118	126	100	642	822	78	3,170
194	79	162	169	100	778	1,029	76	3,703
179	100	180	186	100	643	769	84	3,128
343	100	225	230	100	859	1,010	85	3,475
330	100	211	222	100	788	1,088	72	4,247
276	92	233	211	100	809	1,117	72	4,083
356	100	221	222	100	848	1,129	75	2,983
		295						3,344
		233						2,684
		221						3,668

later company was formally independent. Only the share of the Bengal Nagpore Coal Company varied strongly and averaged only 11%.

But the Katras-Jherriah Coal Company did not only buy from companies under the management of Andrew Yule & Co. In times of rising demand it secured its additional demand through purchases from small companies. The raising capacity of the group itself was obviously oriented towards a relatively stable level of "minimum demand". Fluctuations were therefore shifted to outside companies. These small companies had to bear the brunt of market fluctuations. If one looks at the list of coal mines worked in Bengal one finds indeed a large number of pits and quarries closed in one year and reopened after a few years. These small companies were obviously dependent on the large ones because they did not have any direct access to the market in Calcutta.

In order to assess the raising capacity of the Katras-Jherria Coal Company we have to look into the ratio between the company's raisings to its sales. Was the company's expansion in sales paralleled by an expansion of its production? No, it was not. If we take the colliery expenses of the Katras-Jherriah Coal Company as the valid indicator for its physical production[54] we find that the expenses varied between Rs. 100,000 and 150,000 per year and were not related to the returns from sales at all, i.e. a growing turnover and expansion of sales did not lead to larger units.

In other words, expansion was carried by adding new, formally independent units of production. Thus, the parent company controlled increasing numbers of external companies. These external companies were not financed through direct investments by the parent companies, but external capital was drawn into these ventures by means of selling their shares.

The type of horizontal integration becomes quite apparent. There were key companies controlling the sales and the capacities of formally independent but *de facto* dependent companies. The managing agencies concentrated on these key companies. The strong position of the large companies on the market (access to the market of Calcutta and the export markets) was used to shift the burden of heavy market fluctuations to small independent companies, which only flourished in periods of boom and closed down during recessions.

5.2 *Vertical integration*

Which factor determined the strong position of these key companies on the market? To whom was the coal sold? The sources on which we can draw to answer this question are the sales registers of the Katras-Jherriah Company. Only three of them have survived in the B.C.C. archives, the first from August 1898 to July 1901, the second from February 1906 until July 1908 and the third from February 1915 until January 1917.

These registers contain only the sales invoiced in Calcutta. In the earlier period they contain only varying but relatively large shares of total sales (between 36 and 76% of total sales), for the years 1915 and 1916 they comprise all sales. Tables 7.7, 7.8 and 7.9 give the details of distribution of sales. Table 7.10 demonstrates the high and increasing degree of vertical integration. In the tables only those companies have been regarded as "controlled" by Andrew Yule & Co. which were openly tied to this managing agent. Most probably the actual degree of integration was even higher. Under the heading "owners of steam boats" all of the steamboats

[54]There is no reason to suggest that we should not take them as an indicator because the wages, royalties etc. did not change during the period.

TABLE 7.7
SALES OF THE KATRAS-JHERRIAH GROUP, 1899-1901
(IN % OF ITS TURNOVER IN RS.)

Customers	*Jan. 1899*	*July 1899*	*Jan. 1900*	*July 1900*	*Jan. 1901*	*July 1901*
(1)	(2)	(3)	(4)	(5)	(6)	(7)
Adjai Coal Company+	2,2	—	—	—	—	—
Alliance Jute Mills	—	1,1	3,5	3,8	4,1	3,4
Amulya Nath Roy	0,5	—	2,2	1,0	—	—
Bengal Assam Steam+	4,4	3,1	5,2	3,0	4,0	2,4
Bengal Coal Co.	1.2	—	—	—	0,8	—
Bengal Mills+	11,1	6,6	6,0	3,2	2,4	3,3
Begg, Dunlop & Co.	—	—	—	2,0	—	—
Budge Budge Jute+	19,1	18,5	15,3	17,1	14,6	9,9
Burrakur Coal	—	3,3	—	—	—	2,2
Calcutta Hydraulic	0,8	—	0,8	0,5	1,3	0,7
Cecil, T.S.	—	7,4	5,6	—	—	—
Central Jute Mills+	0,2	—	—	0,1	2,3	—
Steamboats	30,4	21,7	20,1	4,1	5,0	4,0
Delta Jute Mills+	0,4	1,9	4,2	4,2	3,4	4,5
Fort Gloster Jute	3,6	4,1	—	—	—	—
Gillanders Arbutnoth	—	—	—	—	—	1,5
W. Growder & Co.	—	—	—	—	—	2,2
Howrah Flour Mills+	1,7	1,9	1,9	1,6	1,5	1,1
Howrah Jute Mills	1,0	8,0	8,2	10,4	8,5	8,8
Howrah Oil Mills	—	0,3	—	—	—	—
Imperial Paper	—	0,7	—	3,6	—	—
Indian Gas+	0,7	0,9	0,9	1,2	0,9	0,4
Jessop & Co.	2,4	2,7	2,9	3,2	2,4	2,0
Killick Nixon & Co.	—	—	—	1.5	11,2	—
Kunja Behary Ghose	—	—	1,4	1,0	1,2	1,3
Mackinon, Mackenzie & Co.	—	—	—	3,8	4,4	0,4
MacLeod & Co.	—	—	1,2	0,8	2,6	0,8
Monarch Flour Mills+	—	—	—	—	1,6	1,3
Mutty Lal Paul	—	—	1,7	—	0,9	—
National Jute Mills+	4,4	4,7	5,2	4,6	4,3	3,7
Dey & Friends	—	—	—	6,7	11,5	11,4
Steam	—	—	—	9,1	—	25,0
Ralli Bros.	3,4	2,0	2,9	0,8	3,6	—
Miscellaneous	12,1	11,2	10,8	11,4	7,6	9,2
Sum	100	100	100	100	100	100
+ controlled by Andrew Yule & Co.	44,3	42	39,5	35 5	36,3	27,3

Source: Bills Registers of the Katras-Jherriah Coal Company, B.C.C-Archives.

are comprised which transported the Katras-Jherriah's coal to Madras, Bombay and Karachi etc.; it would be reasonable to assume that some or even most of them were somehow incorporated into the vertical integration.

TABLE 7.8
SALES OF THE KATRAS-JHERRIAH GROUP, 1906-1908

Customers	*July 1906 in %*	*Jan. 1907 in %*	*July 1907 in %*	*Jan. 1908 in %*	*July 1908 in %*
	(1)	(2)	(3)	(4)	(5)
Belvedere Jute Mills+	—	—	—	1,2	2,2
Bengal Assam Steamship+	2,8	4,7	2,0	4,2	2,1
Bengal Mills+	5,1	4,2	6.5	7,8	7,6
Budge Budge Jute Mills+	14,9	14,6	13,2	14,9	10,0
Calcutta Hydraulic Press+	0,3	1,9	0,7	1,7	0,7
Central Jute Mills+	0,7	—	0,2	0,3	0,3
Steamboats++	6,0	8,7	6,9	3,0	11,1
Delta Jute Mills+	7,1	7,0	6,8	7,1	6,0
Duff, Thomas & Co.	0,2	—	8,3	10,2	8,9
Fort William Flour Mill+	—	0,9	1,1	0,7	—
Howrah Flour Mills+	2,0	1,8	1,4	2,2	2,8
Howrah Jute Mills	9,9	9,6	8,1	11,3	6,3
Howrah Oil Mills+	3,8	5,4	4,0	—	—
Indian Areating Gas Co.+	0,6	0,2	—	—	—
Jessop & Co.	5,7	5,2	4,8	5,0	3,9
Killick Nixon & Co.	—	3,9	—	—	—
MacNeill & Co.	—	—	—	—	3,9
Miller, R.N.	—	—	2,4	1,0	0,4
Monarch Flour Mills+	1,4	1,7	1,5	0,9	—
National Jute Mills+	5,9	6,0	6,0	6,8	5,4
Ralli Brothers	1,2	3,1	1,0	3,2	1,9
Gun & Shell Factory	4,6	—	—	—	—
N.W. Railway	5,7	1,0	8,3	—	16,0
Seebpore Coal Co.+	—	—	—	1,8	—
Miscellaneous	22,2	19,3	16,7	16,5	10,3
Sum	100	100	100	100	100
+controlled by Andrew Yule & Co.	50,6	57,8	50,3	52,6	48,2

++They were almost exclusively supplied by Andrew Yule & Co. during this period. Most of them were transporting coal.

Source: Bills Registers of the Katras-Jherriah Coal Company. B.C.C.-Archives.

Among the "miscellaneous" companies, a number of companies were tied to Andrew Yule, adding again to the degree of integration.

Another sort of tie existed between the Katras-Jherriah Coal Co and two of its large customers, the Alliance Jute Mill Co. and the Howrah Jute Mills Co., both controlled by Ernsthausen & Co. Andrew Yule & Co. had bought relatively large quantities of their debentures. For example the K.J. Company held debentures of the Howrah Jute Mills Co. worth Rs. 188,500 and the Adjai Coal Company debentures worth Rs. 25,500.

Comparing the tables 7.7-7.9 we see that the number of customers decreased

TABLE 7.9
SALES OF THE KATRAS-JHERRIAH GROUP, 1915-1917
(IN % OF ITS TURNOVER IN RS.)

Customers	*July*[1] *1915*	*Jan. 1916*	*July 1916*	*Jan. 1917*
(1)	(2)	(3)	(4)	(5)
Anglo-Oriental Navig. Co.+	15,5	7,1	—	—
Arracan Co.	4,1	1,7	3,4	—
Belvedere Jute Mills+	—	2,2	2,0	3,7
Bengal Assam Steamship Co.+	3,9	2,4	1,9	3,8
Bengal Coal Co.+	11,5	15,4	17,5	11,1
Bengal Mills Co.+	4,2	3,1	3,1	6,1
Budge Budge Jute Mills+	2,4	0,2	0,1	—
Calcutta Hydraulic Press+	—	0,7	0,4	1,0
Caledonian Jute Mills+	—	—	2,5	0,4
Steamboats	—	—	2,3	—
Delta Jute Mills+	5,7	2,8	3,0	5,7
Graham & Co.	—	2,8	0,1	—
Harperiah Smith & Co.	—	4,6	6,9	—
Howrah Flour Mills+	2,5	1,0	0,7	1,8
Jessop & Co.	0,1	—	—	—
Mackinon, Mackenzie & Co.	1,3	0,6	—	—
Military Departments	1,9	—	2,3	1,4
Murrey Brewery Co.	—	1,2	1,1	1,6
National Jute Mills Co.+	—	2,5	2.9	5,1
New Central Jute Mills Co.+	1,0	2,6	1,8	4,2
Paterson Simons & Co.	—	15,5	11,1	—
Port Engineering Works	—	—	—	1,0
Ralli Brothers	3,9	2,1	1,7	4,6
Railways	22,3	13,5	17,7	33,1
Sinclair Murray & Co.	—	—	0,8	—
Tata Iron & Steel Co.	—	0,2	4,4	1,6
Andrew Yule & Co.	19,7	12,2	8,3	5,8
Miscell.	—	5,4	4,4	7,8
Sum	100	100	100	100
+controlled by Andrew Yule & Co.	66,4	82,2	46,5	48,7

[1]Bills Registers of the Katras-Jherriah Coal Co., B.C.C.-Archives from February to July 1915 are incomplete. Figures are only based on a period of 4 months, e.g. on a turnover of Rs. 670,434.-(altogether Rs. 971,246.—).

Source: Bills Registers of the Katras-Jherriah Coal Co., B.C.C.-Archives.

over the period, the number of larger customers shrank from three dozen to two dozen. Short term customers (whose tenders were usually secured for the period of one year) disappear from the registers. Between 1898 and 1901 there were still quite a number of customers with yearly tenders, e.g. Anglo-Indian Jute Mills, Fort Gloster Jute Mills, Imperial Paper Mills, as well as the P & O Steamship Company. The volume of these deals is

TABLE 7.10
SALES OF THE KATRAS-JHERRIAH GROUP 1898-1916

Financial Year	*Turnover acc. to Bills Registers Calcutta in 1000 Rs.*		*Percentage of sales from Calcutta of the total turnover*		*Percentage of turnover of the Andrew-Yule-Group*
(1)	(2)		(3)		(4)
2/1898	331		62		44
1/1899	452		76		42
2/1899	366		36		40
1/1900	429		52		36
2/1900	500		69		36
1/1901	470		76		27
		average	62	average	37,5
1/1906	579		43		51
2/1906	712		44		58
1/1907	907		57		50
2/1907	782		42		53
1/1908	1,004		52		48
		average	48	average	52
1/1915	971]		100		66
2/1915	1.853]				52
1/1916	2.015]		100		47
2/1916	1.232]				49
		average	100	average	54

much smaller in the years 1906 until 1908. Of the roughly twenty larger customers two thirds belonged to the group of Andrew Yule & Co. themselves.

Up to 1915/17 this tendency had increased even further. But the picture is distorted by the influence of the government's distribution control.[55] During the second half of 1916 33% of the sales went to the railways (this was the average figure for the whole coal industry) and the K.J. Co. had to deliver coal to Tata Iron & Steel Co. Our tables do indeed verify the type of vertical integration which was so often positively mentioned when new companies were established, and we can quantify the degree. Table 7.10 gives the percentage of sales (invoiced in Calcutta), which remained within the managing agents group of companies. This share of sales increased from 37.5% around the turn of the century to 52% in the years

[55]From June 1917 onwards the coal production of all mines raising "first class coal" was requisitioned by Government. In spring 1918 the production of mines with "second class coal" was temporarily placed under requisition. Cf. *Report on the production and consumption of coal in India for the year 1918*, p. 13.

1906 until 1908, to 54% between 1915 and 1916. The strongest expression of this tendency is reflected by the figures for the first half of 1915, before the distribution of coal came under governmental control. During this year two thirds of the sales went to companies controlled by Andrew Yule & Co.

These findings rest on a broadening basis as the percentage of the sales invoiced in Calcutta increased all the time. These sales constituted only 40% of all sales of the K.J. Company around the turn of the century, 48% between 1906 and 1908 and had reached 100% by 1915/16. An additional aspect has been mentioned: a number of external companies were obviously tied to Andrew Yule & Co's coal department by other means than favourable prices or sales conditions, e.g. by credits. The Katras-Jherriah Coal Co., an immensely profitable company with many outlets for profitable investments, bought up debentures worth Rupees 185,000 (an amount equalling two fifth of the K.J. Co's own capital) at an interest rate of 5.5% p.a. It is evident that such an investment was not justified by the debentures' interest rate, but only to secure sales. The same applies to the debenture holdings in the Alliance Jute Mills and the Fort Gloster Jute Mills Co., both managed by Kettlewell Bullen & Co. If we included these "associated" companies into the "integration", the degree would be much higher.

It is quite obvious that the K.J. Co's pattern of coal distribution by industrial sectors was not the average for coal mines in Bengal in the sense of averages. Jute mills were the principal purchasers of coal from the K.J. Co. They bought roughly one third of the K.J. Co's total sales in Bengal, between 1906 and 1908 even 37%. Within each of the large agency groups the pattern of distribution must have had a different structure. It would be wrong to infer from our results that all the managing agencies neglected the market covered by tenders, i.e. especially the railways and the steamer lines. It is well known that the second largest group in the coal industry, Bird & Co. was heavily engaged in this market. But their firm's history, especially the reminiscences of Lord Cable himself,[56] confirm the tendency which becomes apparent in the accounts of Andrew Yule & Co: a decreasing market mobility. All means were employed to establish fixed market relations and assure sales on a long term basis. The market was portioned into compartments. It became impossible for an independent coal mine, if it was not one of the large companies (like the Bengal Coal Co.) to compete with the managing agents.

It is, therefore, misleading to look for competition between individual coal companies. Competition was working on a level of higher concentration, on the level of groups of companies under a uniform control, multi-industrial groups in fact, guided by interests which might have been totally

[56]Harrison, G. '*Bird and Company of Calcutta, 1864-1964*, Calcutta 1964.

alien to one or the other of the industries or companies incorporated in them.

5.3 *Prices and the distribution of profits*

We have found a high degree of horizontal and vertical concentration or integration. Key companies organised the distribution of coal. They bought large quantities of coal from associated and independent companies and sold the coal to associated companies and independent purchasers. It is obvious that the price policy pursued on both these levels—horizontal and vertical—determined the distribution of profits. We shall therefore look into the price policy *vis-a-vis* the customers, (the vertical level) and *vis-a-vis* coal suppliers (the horizontal level), starting with the vertical line.

In analysing the price policy we have to overcome a number of difficulties. The first results from the denominations of qualities of coal, classified in books of the K.J. Co. as hard coke, steam coal, smithy coal and slack coal. 90% of all sales were steam coal, hard coke was sold in minute quantities only. But nevertheless, we shall first look at the sales of hard coke because hard coke was the most clearly defined quality. Table 7.11 shows prices charged for hard coke. Of the eight customers for this type of coke only two did not belong to the group of Andrew Yule & Co.: Jessop & Co. and Ralli Bros., both paid prices which differed clearly from the rest. Whereas all the associated companies paid stable prices (with one exception) until the beginning of 1908, the prices paid by Jessop & Co. moved from 17 Rupees in the first half of 1901 down to 9 Rupees in 1907.

For the group of associated companies we can say that the "internal" prices ignored price fluctuations caused by the business cycle or seasonal ups and downs and changed only in extreme market situations (e.g. in 1908). Secondly, "internal" prices were not frozen at one level, but they were arbitrarily settled for each individual company or industrial branch. Between 1898 and 1901 the Indian Aerating Gas Co. paid Rs. 14 per ton, all jute mills Rs. 12 per ton. Between 1906 and 1908 the Budge Budge Jute Mills paid Rs. 14 all others Rs. 12 per ton.

A look at the price movements for steam coal reveals that these findings are not accidental. All prices in table 7.12 have been recalculated per ton and are expressed in decimals.[57] With regard to "steam coal" we encounter a new difficulty. Dealing with such a vaguely defined quality as "steam coal" it is difficult to build an argument on price differences among various customers because we cannot be quite certain if all purchasers were in fact supp-

[57]All prices were calculated in short tons at 907 kg (this is revealed by price quotations in tons and maunds), one example: the Bengal Assam Steamship Company paid Rs. 6. per ton of steam coal in 1906 or 4 annas and 3 pies per maund, equalling 0.27 Rupees, which is comprised 22.6 times in the price per ton, as often as maunds at 40.1 kg in one short ton.

TABLE 7.11
PRICE POLICY OF THE KATRAS-JHERRIAH COAL COMPANY, 1899-1908
PRICES FOR "HARD COKE" PER TON IN RS.

Customers	*Aug. 1898 Jan. 1899*	*Feb. 1899 July 1899*	*Aug. 1899 Jan. 1900*	*Feb- 1900 July 1900*	*Aug. 1900 Jan. 1901*	*Feb. 1901 July 1901*	*Feb. 1906 July 1906*	*Aug. 1906 Jan. 1907*	*Feb. 1907 July 1907*	*Aug. 1907 Jan. 1908*	*Feb. 1908 July 1908*
(1)	(2)	(3)	(4)	(5)	(6)	(7)	(8)	(9)	(10)	(11)	(12)
Calcutta Hydraulic			12								16
Central Jute	14			12	12		12		12	12	12
Delta Jute	12	12	12	12	12		12	12	12	12	15.50
Indian Aerating Gas[1]	14	14	14	14	14	14	12				
Jessop & Co.			15	14	14	17			9	9	
National Jute	12		12	12	12	12	12	12	12		16
Budge Budge Jute							14	14	14	14	15.50
Ralli Brothers							13.75	13.44 13.75	13.75		
Amount sold in tons	49	138	351	166	129	84	166	174	60	60	246

[1]The Indian Aerating Gas Company only bought about two thirds of "Hard Coke".

lied with the same quality of coal. On the other hand, it seems reasonable to assume that any single customer required a stable quality of coal for his consumption. This means prices ought to have changed always in the same direction and approximately by the same rate. They have been compiled in Table 7.12, most of the purchasers were associated companies.

The main features are: constant prices from 1899 to 1906 followed by two waves of price increases in 1907 and 1908; secondly, the dissolution of the fixed price structure in the years 1916 and 1917. During these two years for the first time seasonal fluctuations were reflected also in deals with associated companies, e.g. with the Bengal Assam Steamship Company.

Let us first look at the price movements within one group of customers, the jute mills. Of the four jute mills belonging to Andrew Yule & Co. three paid constant prices at Rs. 6.10 per ton until the end of 1906, whereas the Budge Budge Jute mills paid a higher price of Rs. 6.78 per ton until the beginning of 1901 and from then onwards the constant price of Rs. 5.88 per ton. An independent jute mill, the Alliance Jute Mills Co., managed by Begg, Dunlop & Co. was charged prices which changed rapidly. In the first half of 1899 the company paid Rs. 4.50 per ton (for three deliveries), from August 1899 until July 1900 Rs. 6.74 per ton (for seven deliveries), from August 1900 to July 1901 Rs. 6.25 per ton (for eight deliveries). The Fort Gloster Jute Mills, managed by Kettlewell and Bullen, paid Rs. 6.78 for the five deliveries which they received from

TABLE 7.12

KATRAS-JHERRIAH COAL COMPANY, SALE-PRICES PER TON OF "STEAM COAL" IN RS. 1898-1917

Customers	*Aug. 1898 -Jan. 1899*	*Feb. 1899 -July 1899*	*Aug. 1899 -Jan. 1900*	*Feb. 1900 -July 1900*	*Aug. 1900 -Jan. 1901*	*Feb. 1901 -July 1901*	*Feb. 1906 -July 1906*	*Aug. 1906 -Jan. 1907*	*Feb. 1907 -July 1907*	*Aug. 1907 -Jan. 1908*	*Feb. 1908 -July 1908*	*Feb. 1915 -July 1915*	*Aug. 1915 -Jan. 1916*	*Feb. 1916 -July 1916*	*Aug. 1916 -Jan. 1917*
(1)	(2)	(3)	(4)	(5)	(6)	(7)	(8)	(9)	(10)	(11)	(12)	(13)	(14)	(15)	(16)
Alliance Jute Mills		4,50	6,74	6,74	6,25	6,25									
Belvedere Jute Mills										8,94	9,94		8,50	8,50	8,50
Bengal Assam Steamship	6,50	6,50	6,50	6,50	6,50	6,50	6,00	6,00	8,25	8,25	9,25	7,75	7,75	7,75	7,50
										9,25				8,50	8,50
														7,50	9,00
Bengal Mills	6,33	6,33	6,33	6,33	6,33	6,33	6,33	7,00	8,56	8,56	9,81	7,00	7,00	7,00	6,75
								6,33						6,75	7,75
								8,56							8,00[2]
Budge Budge Jute Mills.	6,78	6,78	6,78	6,78	6,78	5,88	5.88	5,88	7,94	7,94	9,44	9,75	9,00	7,50	
Steam								6,55[3]		9,44[1]		8,50			
												9,00			
Budge Budge Jute Mills	7,46	7,46	7,46	7,46	7,46	7,46	6,78	6,78	—	—	—				
								7,23[3]							
Calcutta Hydraulic	6,10		6,10	6,10	6,10	6,10	6,10	6,10	8,31	8,31	9,31	6,75	7,25	7,25	7.25
								7,00[3]							
Central Jute Mills				6,33	6,10		5,65		9,75	9,75	11,00		6.75[6]	6,75	6,75
															8,88
Delta Jute Mills	6,10	6,10	6,10	6,10	6,10	6,10	6,10	6,10	8,94	8,94	9,94	9,50	9,50	9,50	9,50
										9,94[4]					8,50
															10,00
Thomas Duff & Co.							6,75		7,25	7.25	8,75				
											6,48[5]				
Howrah Flour Mills	6,25	7,25	7,25	7,00	7.00	7,00	6,75	6,75	7,75	7,75	8,75	7,00	6,75		6,12
		6,25	7,00												

Contd . . .

TABLE 7.12 *Contd.*

(1)	(2)	(3)	(4)	(5)	(6)	(7)	(8)	(9)	(10)	(11)	(12)	(13)	(14)	(15)	(16)
Howrah Jute Mills	5,65	5,65	5,65	5,65	5,65	6,10	5,88	5,88	5,88	5,88	5,88				
Indian Aerating Gas	6,33	6,33	6,33	6,33	6,33	6,33	5,65	5,65	8,57						
Jessop & Co.	6,33	7,00	7,00	7,00	7,00	7,00	8,00		9,00	9,50	7,00				
						8,36					8,00				
National Jute Mills	6,10	6 10	6,10	6,10	6,10	6,10	6,10	6,10	8,94	8,94	9,94			7,63	7,63
										9,94[1]					8,88
Ralli Brothers	6,33	6,33	7,00	7,00	7,00		5,88	5,88	5,88	8,59	8,59				
		6,78													
Steamboats		9,50		9,50	9,50	8,00	8,00	8,50	10.00	9,00	9,00				
		-7,50		-8,50	-8.50		-9,00	-9,00	-10,50	-10.50	-11,00				
N.W.-Railway							3,37	3,38				5.25	5,25	4,75	4,75

[1] from December 1907 onwards.
[2] the two last prices only from February 1917 onwards.
[3] from November 1906 onwards.
[4] from January 1908 onwards.
[5] from June 1908 onwards.
[6] "New Central Jute Mills" from 1915 onwards.

August 1898 until July 1899[58] (same price as Budge Budge).

The lowest price was quoted for Howrah Jute Mills Co. at Rs. 5.65 per ton over five financial years until it was raised to Rs. 6.10 between February and July 1901, at the same time, when the price for the Budge Budge Jute Mills was reduced. It seems, that independent and associated purchasers had a complementary function, when a higher price could be obtained from an independent buyer. The main feature becoming evident in Table 7.12 is: fixed prices for associated purchasers, changing prices for independent customers. This result is supported by the prices quoted for customers in other industrial sectors than the jute industries. They remained stable (in the first period) for the associated companies: Bengal Assam Steamship, Bengal Mills, Calcutta, Hydraulic Press and Indian Aerating Gas. At the same time they fluctuated between Rs. 6.33 and 8.36 per ton for Jessop & Co. Roughly two thirds of the sales were not affected by seasonal price fluctuations.

These fluctuations affected only occasional sales, for example to competing coal companies which had to purchase coal at certain points of time to meet their obligations. Thus, the Bengal Coal Co. paid Rs. 5.50 per ton in October 1898; three months later it could purchase the same quality at Rs. 4.25 per ton. References to seasonal fluctuations are rare, like the remark on an invoice for the Adjai Coal Co. of 31.8.1898, "special price because of plague".[59]

For the first period (summer 1898 to summer 1901) we can answer the three questions rather clearly:

(a) the general price level, reflected in official statistics, did not conform to the price structure of the Katras-Jherriah Coal Co.

(b) there were marked differences between prices quoted for associated and independent purchasers.

(c) the price differential was subject to arbitrary changes.

The period of 1906 to 1908 is even more informative about the character of "internal" price policy. Especially the boom of 1907 and 1908 reveals that associated companies were not charged "prices" in the proper sense of the word, but had accounts with the Katras-Jherriah Coal Company which were clearly manipulated at the end of the financial years. Under the heading "differences in price" we find entries in January and February 1907, relating to deliveries for the last three months and increasing the prices previously paid in an arbitrary fashion.[60] The Bengal Assam

[58]Cf. Bill Nos. 2543, 2601, 2792, 2793, 2804, Bills register, K. J. Co. (B.C.C.-archives).

[59]Cf. Bill no. 2462, Bills register, K.J. Co., (B.C.C.-archives)

[60]Cf. e.g. bill no. 8294, of February 1907.

Steamship Co. was charged an additional Rs. 2.25 per ton, the Fort William Flour Mill Re. 1, the Calcutta Hydraulic Press Rs. 0.94, the Bengal Mills and the Budge Budge Jute Mills Rs. 0.71. The Delta and the National Jute Mills did not have to bear with subsequent price increases, but had to pay even higher prices from February 1907 onwards.[61]

The same type of transactions recommenced in spring 1908, this time the volume of the manipulations was even larger. Some price increases went back to December 1907, others to January 1908, others to March 1908, the increases varied between Re. 1 and Rs. 1.5 per ton.[62] Two aspects reveal that these price increases were by no means necessary measures: the rate of increase differed as well as the periods for which they were charged. And in any case, the Katras-Jherriah Coal Co.'s profit for the financial year (6 months) in question came to Rs. 260,885 and thus there was no need for these increases. It is evident that the price manipulations were the result of an overall profit strategy by Andrew Yule & Co.

Until 1908 price movements within the group of associated companies were relatively uniform, for all companies' prices moved into the same direction, although they were increased by different margins. For the years 1915 to 1917 the pattern is totally different, some customers had to pay increasing prices, for some prices remained constant, in other cases they fell. The largest independent customer, the North-Western Railway Company paid a stable and very low price throughout this period. It seems that this relatively large basis of cheap sales was compensated by a flexible adjustment of prices *vis-a-vis* associated companies with the aim of maintaining the high profits of the Katras Jherriah Coal Co.

In other words, a changed market situation meant that diminishing returns were countered by overproportional price increases for associated buyers. There was thus a tendency to shift profits from other industrial sectors (or coal consuming companies) over to the Katras-Jherriah Coal Co. We shall have to look into the reasons for these shifts. The prices charged to associated companies were high at all times, so this transfer of profits was working throughout the period of analysis.

The price policy on the horizontal level is more difficult to trace. It was the usual practice not to note the unit prices or the quantities of coal purchased from other coal mines. Scattered data on purchase prices suggest that prices were at the lower end of the market. More important however is the practice of rebates, which were fixed at the end of the financial years. So again we find the manipulations of returns according to the needs of the Katras-Jherriah Coal Company or the Managing Agent.

[61]Cf. Bills Registers, Katras-Jherriah Coal Company (B.C.C.-archives)

[62]In January 1907 the subsequent price increases had augmented the income of the Katras-Jherriah Coal Company by Rs. 6,000 only, in spring 1908 they led to a surplus of Rs. 16,500.

Table 7.13

GROSS PROFITS AND DIVIDENDS OF THE COAL MINING ESTABLISHMENT OF ANDREW YULE & COMPANY, 1894-1918

	Adjai Coal Company			Bengal Nagpore Coal Company			Seebpore Coal Company			Minto Coal Company			Katras-Jherriah Coal Company		
Year	*Profits in Rs.*	*Profits in % of the capital*	*Divi-dends in % of the capital*	*Profits in Rs.*	*Profits in % of the capital*	*Divi-dends in % of the capital*	*Profits in Rs.*	*Profits in % of the capital*	*Divi-dends in % of the capital*	*Profits in Rs.*	*Profits in % of the capital*	*Divi-dends in % of the capital*	*Profits in Rs.*	*Profit- in % of the capital*	*Divi-dends in % of the capital*
(1)	(2)	(3)	(4)	(5)	(6)	(7)	(8)	(9)	(10)	(11)	(12)	(13)	(14)	(15)	(16)
1894	10.455	3	2,75												
1895	48.046	14	8,5												
1896	50.385	14	9,0	25.699	11	—							52.856	11	12,5
1897	48.959	14	8,5	22,029	10	—							69.145	14	8,75
1898	60.942	17	10	27.961	12	—							123.192	26	7,75
1899	64.506	18	12	48.608	22	—							160.327	32	18,75
1900	112.748	32	18	109.890	49	15	623						442.734	88	27,5
1901	105.374	30	22	117.841	42	22,5	15.496	6	3,5				242.932	49	35
1902	85.044	24	15	90.194	40	15	30.353	11	7				201.612	40	32,5
1903	75.438	22	13,5	65,391	29	10	45.139	16	7				204.479	41	27,5
1904	83.684	24	13,5	25.763	11	5	60.712	22	10,5				292.420	58	35
1905	84.860	24	15	41.233	18	6,75[1]	63.557	23	10,5				350.460	70	45
1906	111.679	32	18,5	118.987	53	20	91.943	33	19,75				426.392	85	60
1907	183.860	53	30	224.196	99	45	141.755	51	32,25				554.313	111	80
1908	165.940	47	32,5	480.577	214	100	164.640	59	41,5	12.749	4	—	735.718	147	100
1909	91.168	26	19,5	140,836	63	55	123.216	44	32,25	48.853	16	17,5	570.214	114	110
1910	106.589	30	20	119.854	58	30	120.591	43	32,25	61.028	20	20	471.298	94	100

Contd. . .

TABLE 7.13 *Contd* . . .

(1)	(2)	(3)	(4)	(5)	(6)	(7)	(8)	(9)	(10)	(11)	(12)	(13)	(14)	(15)	(16)
1911	81.065	23	16	62.622	28	15	92.651	33	25	79.312	27	35	521.778	104	95
1912	83.895	24	16	47.929[1]	43	22,5[2]	190.364	68	44	118.338	40	24,5	508.792	102	95
1913	86.907	25	17			35	178.650	64	49,5	102.815	34	25	576,887	115	95
1914	113.248	32	17			35	170.772	61	50,5	111.497	37	25	622.006	124	90
1915	122.577	35	18			40	177.158	63	48,25	93.627	31	25	583.908	117	90
1916	143.805	41	19,5			32,5	127.501	45	29	128.351	43	27,5	547.293	109	90
1917	91.218	26					44.204	16	9	76.391	26	17	531.276	106	90
1918+	115.662	33					151.220	54	37	73.012	24	14	411.070[1]	164	
Average		27	16,25		47	24		40	27		27	21		84	61
Dividends paid (in %)		60 %			51 %			68 %			78 %			73 %	

[1]First part of fiscal year only. [2]Taken from *Investor's India Yearbook* 1917 from 1912 onwards, Calcutta 1918, p. 146.
Source: Journals from the different firms. B.C.C.-Archives.

5.4 *Analysis of profits*

The American Institute of Management has developed a guideline for assessment of the quality of management.[63] One of the most important point is the so-called "health of earnings". There would not be any way of applying most of the criteria developed by this institute to the managing agencies, but judging by the "health of earnings", the management must have been of an extraordinary quality. The profits gained by the Katras-Jherriah Coal Company were extremely high.

First of all we shall try to find out the actual level of profits, i.e. the profit before depreciation which provided the basis for the commission paid to the managing agent. Secondly, we will analyse the assumption that profits tended to be levelled out by the managing agents. In other words, we will try to find out the profit-differential between the coal companies of Andrew Yule & Company. And finally, were profits in the coal mining department of Andrew Yule & Company higher than those of companies in other industries and were they higher than profits made by other coal companies, managed by other agents? And, were high profits in coal mining the result of transactions from other industries and what were the reasons for such transactions? Table 7.13 shows the profits of the Katras-Jherriah Coal Company and its dependent companies for the whole period of analysis. These profits were shown as "net profits" at the end of each financial year.

Whenever one finds extremely high profits, questions are being raised on the reliability of data. Were the accounts manipulated? Were there different types of accounts? Were the companies undercapitalised? As far as profits are concerned the sources are the most reliable we could hope for. All parties concerned were paid on the basis of the one decisive entry in the half-yearly journals "revenue to sundries". The shareholders received their dividends on this basis, the managing agents their commission (in the cases we are dealing with always 10% of net profits before depreciation) and the manager in the mines their commission (of 5%). No positions were incorporated into these calculations which left room for manipulations, like depreciation etc., so the data are not only reliable but extremely clear.

In order to make profits comparable we have to look at the capital invested into the companies concerned. The Adjai Coal Company had an authorised and paid-in capital of Rs. 3,50,000, the Katras-Jherriah Coal Company of Rs. 5,00,000, the Minto Coal Company of Rs. 3,00,000, the Seebpore Coal Company had an authorised capital of Rs. 4,00,000,

[63]Lloyd K. Marquis: 'A comprehensive framework for analysing the management of a business enterprise', in: *Papers of the Sixteenth Business History Conference*, February 21-22, 1969. Ed. by Charles F. Kennedy, Lincoln 1969, pp. 38-48.

of which only Rs. 7 per share or the sum of Rs. 2,80,430 had been paid in. Only the Bengal Nagpore Coal Company knew the distinction between ordinary shares (Rs. 2,25,000) and perference share (Rs. 1,12,500).

On an average over the years until 1918 the companies earned profits at the following rates:

TABLE 7.14
AVERAGE RATES OF PROFITS P.A. OF FIVE COAL COMPANIES UNDER THE MANAGEMENT OF ANDREW YULE AND CO. (FROM THE YEAR OF THEIR LAUNCHING UNTIL 1918)

Name of company	*rate p.a.*
Katras-Jherriah	84%
Bengal Nagpore	47%
Seebpore	40%
Minto	27%
Adjai	27%

The two offsprings of the Bengal Nagpore Coal Company ranged below the Adjai Coal Company. The Bengal Nagpore Coal Company is the exception in this group. Its earnings fluctuated strongly. For reasons mentioned elsewhere the company was the object of heavy speculation. Between the Katras-Jherriah Coal Company and the three coal companies totally depending on it we find a marked differential of profits. The profit rate of the Seebpore Coal Company was less than half that of the Katras-Jherriah Coal Company, the rate of profits reached by the Minto and the Adjai Coal Co. never surpassed 30% of the Katras-Jherriah's rate.

Throughout the period the role of the companies in this hierarchy did not change much. The share of the Katras-Jherriah Coal Company in the profits of the whole group never fell below 50% and often reached two thirds. The difference between its position and that of the Adjai Coal Company is quite obvious. The Adjai Coal Company's profits remained almost constant (in absolute terms) throughout the period, its share in the group's profits however sank from 50% in the beginning to 12%. This company's shareholders must have been people outside the firm of Andrew Yule.

The special role of the Katras-Jherriah Coal Company on the other hand is not only revealed by the fact that this company always took the lion's share, but that it took an amount which surpassed its share capital every year irrespective of the accumulated profits of the group. In 1917, e.g. the accumulated profits of the group fell from Rs. 9,50,000 to Rs. 7,43,000, the profit of the Katras-Jherriah Coal Company on the basis of the previous year's share (58%) should have sunk to Rs. 4,29,000 Instead the Katras-Jherriah Coal Company took Rs. 5,21,000 or 71% of the accumulated profits.

From these results we have to draw two conclusions:

(1) there was a marked hierarchy of companies, expressing itself in large differences of profits,

(2) the only reason for these differences could lie in varying degrees of capital interest of the managing agents themselves in certain companies.

For if the managing agents had taken their earnings mainly from the commissions it would not have mattered in the least if the profits were concentrated in one company or distributed evenly over all companies concerned.

In order to compare the profits of the Katras-Jherriah group we have to take the average earnings of this group of four companies. Over the period under investigation the group showed an average profit of 48% p.a. (on the accumulated share capital), of which 70% were paid out, resulting in an average dividend of 33.5%. Even this result, much lower than the peak, which was concentrated in the Katras-Jherriah Coal Company, was not achieved in any other industry, and no other industrial sector showed the same stability of the profits. Extraordinary profits of this kind were usually never achieved in times of peace even within the business empire of Andrew Yule & Company. Profits of this volume were only known in times of war, famines and the like.

Finally we have to ascertain if profits were shifted from other industries into the coal mining sector. Again we shall take the average profit of the group of four companies as the basis. Were the group's profits exceptionally large for the coal mining sector? Were these large profits the result of transfers of profits (*via* prices) from jute mills, steamer companies etc? There are only few independent companies, which we can take as control group. The Equitable Coal Company provides a good example. It was an old established company, it reached a dividend of more than 100% p.a. in one year (1904) and ranked among the highly successful coal companies.

The Equitable Coal Company's profits fluctuated different to the ones of the Katras-Jherriah Coal Company and its group, but nevertheless, over the period from 1900 to 1916 its profits averaged 47%. That means an independent coal company could reach profits of this volume too. Certainly there was a difference between an old-established company which had accumulated reserves over 40 years before 1900 and could display its earning capacity in the following years and a company which had only just been established. Nevertheless, the result suggests that the manipulations of the managing agent were aimed more at stabilising earnings in the coal mining industry (on a very high level) than transferring profits from other industries and thus tip the "terms of trade" between the different industries under his control in favour of the coal mining industry far beyond the terms existing "on the market".

5.5 *Investment policy*

Above we have listed the group of companies in the sequence of their profits (before depreciation and commissions). This hierarchy remains the same when we look at the dividends, these companies paid out. On an average they were:

TABLE 7.15
AVERAGE DIVIDENDS P.A. OF FIVE COAL COMPANIES UNDER THE MANAGEMENT OF ANDREW YULE AND COMPANY (FROM THE YEAR OF THEIR LAUNCHING UNTIL 1918)

Name of company	*Dividend in % p.a.*	*% of profit paid out*
Katras-Jherriah	61	73
Seebpore	27	68
Bengal Nagpore	24	51
Minto	21	78
Adjai	16.5	60

On an average the Seebpore Coal Company paid out a higher dividend than the Bengal Nagpore Coal Company. This difference is explained by the ratio of profits paid out. The Bengal Nagpore Coal Company paid out roughly 50% of all profits earned, Seebpore 68%, the Katras Jherriah 73% the Minto Coal Company even 78%. The averages of course level out very dynamic developments. The Katras-Jherriah Coal Company e.g. reached a level of 50% dividend in 1906, between 1908 and 1910, of 100% and this dividend did not drop below 90% from 1911 onwards, so the average is kept down by the first four years with dividends below 20%.

In order to establish the quality of such dividends we have to compare them with interest rates on the capital market in Calcutta. Debentures might give us a lead. Between 1900 and 1910 debentures were offered on the market at 5% p. a. (10 to 20 years) and were raised only once by 0.5% at the beginning of the war in 1914. Preference shares (with a greater risk) were offered at 7%. The actual interest rate was slightly below 7%. Andrew Yule's finance department charged the Katras-Jherrih Coal Company 7% interest for short term credits.

Finally, if we take the price of shares for the Katras-Jherriah Coal Company and compare it with the dividends, we find that the interest rate ranged between 11 and 12%. In 1908 e.g. the price of the Katras-Jherriah Coal Company's shares at the nominal value of Rs. 10 ranged between

Rs. 112 and 90. Taking the average at Rs. 100, the interest rate was 12%. In later years (when dividends dropped to 90%) the price of shares dropped to Rs. 80 and the interest rate to 11%.

For those shares which were bought when the Katras-Jherriah Coal Company was established, or shortly afterwards, the rate of interest was nine times as high as on relatively good purchases of shares, twenty times higher than interest on debentures and thirty times higher than the usual rate of interest on savings (expressed e.g. by the 3% papers issued by the Government of India just before the war).

While analysing the profits of the Katras-Jherriah group we compared the average profit of 48% with the 47% profit of the independent Equitable Coal Company. This company, however, paid out an average dividend of 24% only (bonus of 100%, paid out in 1904, was not lost to the company, but remained in the company). In other words, the company paid out roughly 50% of its profits.

In contrast to this policy, the Katras-Jherriah group followed a policy of vigorous consumption of the company's resources. On an average, 70% of the profits were paid out, there was not much to be put to reserves. The first secret of success was the policy of a profit hierarchy, the second one the policy of vigorous consumption. This policy could be followed, because the managing agent was able to mobilise financial resources for short periods at any time.

Why was the Katras-Jherriah group able to pay out so much more than the Equitable Coal Company? This question leads to the problem of depreciation and investment. Was there anything like a common practice in writing off depreciation in Indian coal mining? No, there was nothing of the sort. The practice was to write off round sums according to the takings of the financial year or to write off the total investments of the year.

Table 7.16 shows the depreciation written off and the investments of the five coal companies within the Coal Department of Andrew Yule & Co. we have been dealing with. In the table, the initial investments have not been considered. In most cases all costs of running the coal mines for the first two years were comprised within the initial investment. In our context only one aspect is of interest: was the capital stock augmented by reinvestments of parts of the profits? Was the potential for raising coal increased or decreased? Was the capital accumulated in this industry geared towards helping this industry to grow? To find out what happened we have to sum up the amounts of depreciation and those of investments and gather from the differences if in the process capital was accumulated or consumed.[64]

We shall start with the Adjai Coal Company. Originally Rs. 3,19,912 were invested in buying the ground and equipping the mines. From 1895 until 1918 Rs. 5,44,353 were written off. During the same period the company invested only Rs. 2,17,096, i.e. 12% of the profits. The net disinvestment amounted to more than Rs. 3,27,000, i.e. more than the initial outlay. Of course, the company had bought securities to cover the original input of capital, debentures, government certificates and the like, but they had nothing to do with the running of a coal mine. In the case of the Bengal Nagpore Coal Company the disinvestment was of a similar volume—within 17 years (up to 1912) almost Rs. 3,50,000 more than the ordinary share capital of the company. Still, 28% of the profits were reinvested, more than double the rate of investment of the Adjai Coal Company.

The Seebpore Coal Company seems to be an exception. After 18 years the disinvestment was only 10% of the original investment. This was caused by relatively large investments during the war years. More than half of the accumulated investments of this company (Rs. 3,85,000 in 18 years) were paid out between spring 1916 and summer 1918. This was the Seebpore Coal Company's share in an electric power plant paid for by the Katras-Jherriah, the Minto and the Seebpore companies. The Minto Coal Company shows a similar policy. In the course of 11 years Rs. 2,33,000 were written off as depreciation, Rs. 1,37,000 or 14% of the profits were reinvested.

The Katras-Jherriah Coal Company shows the lowest rate of investment. Over the period of 23 years only 8% of the profits were reinvested. The largest part during the first years, later only 2 or 3% p.a. During the period under investigation Rs. 9,90,000 were written off, Rs. 8,02,000 were invested, the disinvestment amounted to Rs. 1,79,000 or 36% of its share capital.

Comparing depreciation and investment we have to take into account the material basis for depreciations. Were they very large and were they used for accumulating hidden reserves, or did they just cover the actual loss in value of the machines? Let us take the original investments, calculate the sum of depreciations over the years and transform this into linear depreciation p.a. In the case of the Adjai Coal Company the average depreciation p.a. amounted to Rs. 22,700, i.e. 7% of the original capital

[64]To ensure that such a calculation leads to any valid results it is important to establish if the depreciation was realistic or if high rates of depreciation led to hidden reserves. In the case of the Adjai Coal Company the total amount of depreciation over the period of investigation divided by the number of years was Rs. 22,700. p.a., equalling a linear rate of depreciation of 7% p a. In other words: the assumption was that the equipment would last 14 years on an average: no rate to build up hidden reserves.

TABLE 7.16

INVESTMENTS AND DEPRECIATIONS—COAL MINING ESTABLISHMENT OF ANDREW YULE AND COMPANY, 1894-1918

Year	Adjai Coal Company			Bengal Nagpore Coal Co.			Seebpore Coal Co.			Minto Coal Co.			Katras-Jherriah Coal Co.		
	Depreciations in Rs.	Investments in Rs.	Investments in % of the gross profits	Depreciations in Rs.	Investments in Rs.	Investments in % of the gross profits	Depreciations in Rs.	Investments in Rs.	Investments in % of the gross profits	Depreciations in Rs.	Investments in Rs.	Investments in % of the gross profits	Deprecia- in Rs.	Investments in Rs.	Investments in % of the gross profits
(1)	(2)	(3)	(4)	(5)	(6)	(7)	(8)	(9)	(10)	(11)	(12)	(13)	(14)	(15)	(16)
1894		(319.912)													
1895	6.000	16.963	35											(63.109)	
1896	12.000	11.028	22	22.590	37.278	145							25.000	41.204	78
1897	18.000	12.860	26	21.993	26.809	122							25.000	56.026	81
1898	18.000	4.676	8	25.717	16.802	60							50.340	20.232	16
1899	14.000	6.817	11	44.670	8.567	18							38.506	32.165	20
1900	22.000	2.485	2	66.385	25.610	23	—	(10.333)					40.000	16.356	4
1901	22.000	20.616	20	57.375	18.398	13	—	25.909	167				44.064	12.002	5
1902	18.000	3.621	4	46.581	13.491	15	8.438	15.485	51				36.000	7.240	4
1903	18.000	925	1	41.023	4.712	7	20.566	3.925	9				39.936	5.034	2
1904	20.000	6.247	7	13.565	8.572	33	26.787	4.171	7				55.000	4.776	2
1905	20.000	4.958	6	17.208	20.863	51	26.636	4.554	7				55.000	7.091	2
1906	27.000	3.481	3	49.196	23.105	19	27.014	20.373	22				76.311	26.831	6
1907	50.000	21.155	12	96.316	38.223	17	36.943	3.973	3				69.891	76.014	14
1908	31.000	17.790	11	136.292	77.189	16	35.268	12.632	8	10.168	12.258	96	72.677	56.266	8
1909	10.500	10.640	12	34.555	34.116	24	18.884	19.848	16	18.117	14.582	30	38.491	18.228	3
1910	20.000	10.226	10	57.983	43.969	37	21.509	27.572	23	22.000	16.015	26	14.979	13.163	3

Contd . . .

TABLE 7.16 *Contd*. . .

(1)	(2)	(3)	(4)	(5)	(6)	(7)	(8)	(9)	(10)	(11)	(12)	(13)	(14)	(15)	(16)
1911	13.500	5.146	6	33.986	37.581	60	12.460	9.845	11	18.028	11.787	15	9.864	10.016	2
1912	15.000	500	1+	26.886	15.111	32	42.772	12.434	7	34.264	652	1	5.769	8.297	2
1913	15.735	5.735	7				21.805	6.827	4	16.922	3.922	4	45.445	22.984	4
1914	36.235	457	0				11.807	1.735	1	25.000	—	0	61.798	61.838	10
1915	40.449	14.283	12				20.000	—	0	10.000	—	0	78.265	78.265	13
1916	54.259	17.795	10				33.081	48.051	38	33.596	18.596	14	56.446	56.446	10
1917	15.692	15.692	17				15.569	92.569	209	18.533	3.533	46	20.135	87.621	16
1918+	27.893	3.000	3				32.057	63.057	42	26.556	55.566	76	28.242	89.684+	22
Sums	544.353	217.096	ϕ^{+++} 12	792.621	447.396	ϕ^{+++} 28	411.566	383.293	ϕ^{+++} 23	233.194	136.911	ϕ^{+++} 14	987.079	807.779	ϕ^{+++} 8
Disinvestment		327.257			345.225			28.273			96.283			179.300	

+Only first six months.

++This position was not included in the capital stock, originally not included under Investments.

+++The average did not take into consideration the first two years, as in the initial phase mere working expenses were included under "investments" to diminish losses.

outlay. This means the equipment was meant to work 14 years. This rate is not large but rather small. The Seebpore Coal Company's rate was 7.7%, the Minto Coal Company's 7.1% and the Katras-Jherriah's 8.2% p.a. . In other words, the practice was not geared towards accumulating hidden reserves.

Altogether we can state that there was nothing like an investment policy. It is almost incredible that a company taking Rs. 3.5 million p.a. from sales (like the Katras-Jherriah Coal Company in 1912) invested but Rs. 9,000 during this year.

The analysis of these small investments by different categories reveals that most of the expenditure was for maintenance. Under "machinery" we find entries for spare parts, tools, one pump or another, but with the exception of the war years never any investment into the modernisation or opening of a new shaft, never any investment into underground transport, mechanisation of raising, ventilation etc. The Katras-Jherriah Coal Company's "investments" consisted mainly of expenditure for "new works", i.e. the preparation of new coal faces. The analysis of the cashbooks reveals that the same sums invested could only just keep up the absolute minimum of maintenance.[65]

5.6 *Dividends*

It has become obvious that profits were not reinvested into the coal mining industry. What happended to them? They were paid out as dividends. What, then, happened to the dividends? We do not know. The little we can reconstruct is this: only a minute minority (1%) of shareholders of the Katras-Jherriah Coal Company[66] were resident in Britain and presumably transferred their dividends directly to Europe. Roughly 8% of shares were held by Indian shareholders. Their dividends might have been reinvested into Indian industries, their behaviour in collecting the dividends (some of the larger ones took months to send in their warrants or sometimes forgot to collect the dividends altogether) would not suggest that they were keen investors.

11 to 12% of the shares were held by British investors resident in India, most of them obviously interested in speculation, because they did not hold their shares for much longer than two or three years. It is impossible to state with any precision what they did with their dividends.

80% of the shares were held by banks and managing agency houses.

[65]Even in 1908 when the Katras-Jherriah Coal Company had record profits and invested more than in other years the amount invested for machinery reached only Rs. 33,604. On an average the K.J. Co. invested 52% for "new works" and only 26% into machinery and equipment.

[66]The only one for which we have the list of dividends paid out from the thirty-first to the forty-fifth dividend, 1908 until 1915.

Again, it is impossible to say, if their dividends went into further Indian investments. Indeed, we do not even know, if the banks and agents were the ultimate shareholders or if they only held deposits for shareholders.

In the context of this short paper we cannot go into all the details of the share-holding policy of Andrew Yule & Co. One result of a more detailed analysis[67] was that we do not only have to take into account the shareholding and the interest of the managing agent in concentrating dividends to the key companies, but that we have to consider the speculation of managing agents themselves. Andrew Yule & Co's shareholding in the Katras-Jherriah Coal Company, in any case, dropped during the years when the shares were being sold at ten times the nominal value and the profits of the managing agent from selling these shares seems to have been higher than that gained through dividends.[68]

Altogether, the growing international involvement of the managing agency house Andrew Yule & Co. would suggest that capital accumulated in their coal companies was not only drawn out of the industry from where it originated, but also from Bengal and India.

6. *Coal mining as an example of stunted growth in a colonial economy*

Under the heading of development aid the industrial nations today export three things: capital, technology and management. To a large extent the lack of these three factors accounts for underdevelopment. The managing agencies, the organisers of industrialisation in Bengal in the colonial setting did not lack any of them. Capital was available, technical know-how relatively easy to buy, modern management was their business (if we follow Chandler's[69] definition of centralised purchasing, centralised planning of production by branches and centralised distribution). Nevertheless, the system did not lead to sustained, modern growth.

What, then, hindered a development along modern lines? The parameters for decision-making in Indian coal mining were the same as anywhere else: geological conditions, the legal system of ownership of mining rights, the availability of labour, the infrastructure of distribution and the market conditions. Did any of these factors or a combination

[67] A detailed analysis of shareholdings and an estimate of the income of the managing agency house Andrew Yule & Co. is contained in the author's dissertation pp. 293-310.

[68] The analysis of Andrew Yule & Company's income from their investment into the Katras-Jherriah Coal Company leads to the conclusion that they had invested in 1900 Rs. 4,55,750 and gained from this investment up to 1916 Rs. 51,20,142 about 25% of this income stemmed from agents' commission and allowances, 33% from dividends and 42% from sales of shares.

[69] Cf. Alfred D. Chandler Jr.: *Strategy and Structure. Chapters in the History of the Industrial Enterprise.* Cambridge, MIT, 1969 (1962), p. 24.

of them present a major obstacle to a modern development and intensive growth?

It does not seem so: (a) geological conditions were extremely favourable and invited mechanisation; (b) the lack of mining legislation and the undecided question of mining rights through most of the nineteenth century might have helped chaotic forms of exploitation of mines but certainly was not responsible for it; (c) there was neither a lack nor a surplus of labour. Labour was available irregularly (seasonal fluctuations), but there was never a longterm shortage. It is impossible to argue that a lack of labour hindered growth or the surplus of labour prevented labour-saving innovations. It would rather seem that within the colonial setting the coal mining companies had their ways and means of tying miners down and forcing them to work for them; (d) did the infrastructure of transport prevent modern development? It is certain that the railways were not equipped to handle the transport of coal in a fashion favourable to expansion. Neither were the harbours equipped for handling expanding exports. But then, the railways were able to cope with the growing quantities of coal in one way or another and the low stocks of coal which the coal companies carried suggest that the transport problem could not have been too difficult an obstacle; (e) were the market limitations the main obstacle? Certainly, the lack of an Indian iron industry weighed heavily. But we cannot overlook the fact that export markets provided chances for expansion which were used by other countries but not by the Indian coal mining industry.

The growth of the Indian coal mining industry which we have analysed up to the end of World War I was not of an intensive but of an extensive type. Not only that the technology did not undergo any decisive changes, the units of production did not grow in size either. On the contrary, they seem to have shrunk over the period. Mines of similar size with a similar type of technical equipment were added to each other. The same was happening with the companies themselves. Only a few expanded by means of their own capital. The Indian capital invested in coal mining grew because the number of coal mining companies increased, but the average capital outlay of Indian coal companies sank during the period of investigation.

The explanation for this peculiar type of growth, which lacked all the modernising features of capitalist industrialisation, resulted to a large extent from the system of colonial capital organisation under the managing agents. Whereas independent companies followed the path of intensive growth, relatively heavy investments into the modernisation of mines and increasing productivity (the Bengal Coal Company is an outstanding case throughout the time until 1908, when it was incorporated into the house of Andrew Yule & Co),[70] because they did not have any other chance to compete with others on the open market, the groups of companies under the management of large agency houses seemed to be even more successful. Companies like the Bengal Coal Co., were extremely profitable by any European standards. They were, however, not successful enough in the colonial setting. Their policy of building up reserves, of using the resources relatively carefully, could not compete with the policy of vigorous consumption, exploitation and waste of resources followed by the managing agents.

It was their policy of gearing profits into certain companies, setting examples of incredibly profitable enterprises which lured investors to invest their capital into the new ventures of successful agency houses. Through the means of binding these companies to their management, of speculating with the shares of these companies, of arbitrarily gearing the profits to key companies, wherever it was most profitable for the managing agent, these managing agency houses grew at an incredibly fast rate, controlling industrial empires without investing in them and without making any innovative contributions.

The process of horizontal and vertical concentration which was the motor of capitalist industrialisation in Europe, North America or Japan, did not have the innovative effects on the production under the managing agency system in the colonial setting. This system did not have an economically sound basis, but certainly was a most flexible and versatile system of exploitation and control, without any hesitation to manipulate financial operations through means of price policy or dividends.

We have been dealing with one of the two most successful managing agency houses in Bengal. Do the results justify general conclusions? Yes, they do. Within the house of Andrew Yule & Co. we find the same features of development typical for all industrial sectors in Bengal. Other agency houses controlled smaller numbers of mines or factories. Their interests might have been confined to only some of the typical industries of Bengal. Their place within the network of colonial interests was less important, still they employed the same strategy within the system.

The managing agency houses, it seems, had a rather important function in the 1890s. Within a few years they organised a relatively integrated growth of a number of industries in Bengal. They had the access to the capital market, they had the knowledge of the markets. Without their knowledge of the markets for bunker coal in India, Ceylon, Burma and Singapore the coal mining industry of Bengal might not have experienced such a major breakthrough.

[70]Chapter 9 of the author's dissertation deals with the behaviour of the Bengal Coal Co, tabulated data show the structure of this company's investments from 1890 to 1906. Investments totalled 39 lakhs of Rupees, 30% for new work, 22% for machinery 17% for "plant in use", 27% for purchase of land.

The integrative function they had in the very early phase carried in itself the seed for the perversion of this system to one which did not need intensive growth and development to realise profits, but got along with extensive growth and finally stagnation (in the interwar years). The managing agents formed groups of companies, competing against each other not as industries but as financial groups. Under the conditions of largely integrated distribution and harmonised transactions under their control, they concentrated on building strategic positions within the share capital of the groups they controlled. There was a growing contradiction between their complicated financial management and their outdated modes of production. Thus within a very short period of time the managing agency system developed purely parasitical features and functions.

Dietmar Rothermund

Conclusions

THE origins of economic underdevelopment of the coalfield district are clearly delineated in the historical studies which have been presented here. In the rural sector a quasi-feudal system of the control of land and labour acted as an impediment to agricultural productivity. This system was strengthened as well as distorted by the impact of British colonial rule. The zamindars, the superior landlords of the area, were neither improving landlords who promoted agriculture nor did they function as "dual" landlords who invest elsewhere what they extract from agriculture. Their income was squandered by means of conspicuous consumption or wasted on a host of parasitic retainers. Most of the estates were encumbered. Intermediaries and moneylenders were the real beneficiaries of the wealth derived from agriculture. The peasantry which had become a tenantry under the impact of the British legal system was protected by legislation which emphasized administrative tutelage that rarely benefited the actual cultivator of the soil. Numerous intermediaries knew better how to make use of the law so as to protect their interests.

The peasants eked out a living at the subsistence level and when they sank below this level they sold themselves to coolie recruiters. The drain of manpower by coolie migration meant that less labour was employed locally for land reclamation and melioration. This gave rise to a vicious circle. Agriculture stagnated and could not provide enough food for the people, therefore those who were poor and able-bodied left the area and did not participate in increasing its productive potential. When coal mining was started in the area this only meant an additional demand for coolie labour which fitted into this pattern.

The quick growth of the mining industry led to an influx of labour from other regions because the limits of regional labour supply were reached very soon. The urban enclave which came into existence in this way did not stimulate the social and economic development of the region. Besides what may be termed the coolie linkage no other linkages emerged between the new urban centre and its immediate hinterland. This pattern has shown a remarkable persistence over several decades. As the agro-economic studies show, the area is characterized by a fragmentation of markets for agricultural produce, the isolated local market remaining unrelated to the urban market even at present. Accordingly there is not much of a response to innovations in agricultural production, and capital formation in terms of land reclamation or melioration remains at a low level.[1]

The urban enclave which emerged under colonial rule was controlled by entrepreneurs who had no interest in developing the area and who, in fact, disinvested more than they invested in it. The British managing agency system which dominated the coalfield was well suited to this mode of operation. Coal mining was undertaken only in order to satisfy the limited demand of the export-oriented industries controlled by the same managing agency network. There was no free market, and local entrepreneurs would have had no chance to organize the production and supply of coal on their own. The demand-centered organization of the mining industry thus effectively excluded the possibility of the rise of local entrepreneurs. As the studies of the economists have shown this heritage was crucial for the development of the area even after independence. The Indian entrepreneurs who took over the mining business from their British predecessors came from other parts of India where they had had better opportunities of gaining experience in international and inter-regional trade than the people of Bihar.[2]

If we look at the economic backwardness of the coalfield district in historical perspective this perspective is very different from that of Alexander Gerschenkron who has used these terms in the context of European economic history.[3] To Gerschenkron backwardness means a time lag between the emergence of innovations in one country and their adoption by another. This time lag has its own benefits because it permits the adopter to profit from the experience of the innovator and to fit the innovations into the context of his own economic system so as to make the best

[1]For these studies see D. Rothermund, E. Kropp, and G. Dienemann (ed.), *Urban Growth and Rural Stagnation*, New Delhi 1979.

[2]Ibid.

[3]A. Gerschenkron, *Economic Backwardness in Historical Perspective*, Cambridge, Mass.: 1962.

use of them. In doing this the adopter may even surpass the innovator and thus convert his erstwhile backwardness from a liability into an asset.

But this can be achieved only if the adopting system can freely respond as a whole to the respective innovation, whereas the isolated imposition of innovations leads to the formation of enclaves and a distorted development which does not help to overcome economic backwardness. One of Gerschenkron's examples for the way in which backwardness was turned into an asset is the growth of modern industry in continental Europe where the English precedent was followed on a much larger scale and where big investment banks financed this industrial expansion while in England both technology and industrial finance had much more modest dimensions. If we compare Gerschenkron's example with the way in which a very restricted industrialization was introduced by the British in India and how it was controlled through the managing agency system we immediately see economic backwardness in that different historical perspective to which we have referred above. It is a perspective of an economic backwardness which remains a liability and cannot be turned into an asset, because there is no response of an adopting system as a whole, but only isolated impositions and fragmented responses which do not contribute to economic development. In fact, some of these fragmented responses like the conversion of feudal control into landlordism or the coolie linkage mentioned above are retrograde and impede economic development. Backwardness is thus enhanced rather than overcome by the adoption of such innovations.

The field studies of the contemporary situation show that this negative heritage which has been outlined here in historical perspective still encumbers economic development. This negative heritage cannot be redeemed easily, it pervades the structure of contemporary social and economic life.

Glossary

abwab	levy in kind or cash in addition to rent
ahar	local irrigation works
anchal	block, next administrative unit to the district
anna	old coin, there were 16 annas to one Rupee
bakasht	land at the disposal of the landlord which was cultivated either with servants, forced or bonded labour or share croppers
baku	rent paid in kind in terms of half a share of the crop
bandh	dam or embankment of irrigated fields
beth begari	forced labour on the landlords' fields
beth keta/kheta	land under service tenure
bhugut bandha	usufructuary mortgage for a specified period
bhuinhar	(Oraon) male descendant of the original village founder, by custom he paid no rent or only a nominal one
bujharat	Plot to plot enquiries made to prepare land records at the time of the abolition of zamindari system
chanda	village contribution to the expenses of the raja, originally assessed by mutual agreement
diku	stranger, non-adivasi
don	low (rice) land
ijardar	permanent tenure-holder, village headman

kamia	bonded labourer
kamiauti	bonded labour
khas raiyat	tenant who holds his land directly from the landlord
khunt	(Munda) family branch of a clan
khuntkattidar	(Munda) male descendant of the village founder, holding a privileged position within the village
kili	(Munda) clan
kismat	share of property, sub-division of land among several heirs
korkar	newly reclaimed rice land, generally rent free for some years, and then assessed at half the usual rent rate
majhas	see *manjihas*
manjihas	same as bakasht, but surveyed and registered under the Chota Nagpur Tenure Act of 1869
Manki	(Munda) headman of 10-12 villages
mukkarrari	permanent tenure with unalterable quit-rent
Munda	(Munda) secular village head, originally subordinate to the *Pahanr*
Mundari khunkattidar	see *khuntkattidar*
nagdi abwab	levy paid in cash when the rent receipt was issued
Pahanr	(Munda) religious village head
paila	a pot used for measurement
parha	a group of 10-12 villages
parja	collateral relative of village founder
pati	see *parha*
rajhas, rajhus	land assessed with regular, full rent rates
rakumat	same as abwab
saika	rent paid in kind in terms of a fixed amount of grain
salami	an entrance fee levied from a new tenant at the beginning of his tenancy or a gratuity or offering on receiving a lease
sardar	leader
Sardar Larai	War of the Leaders
tanr	up-land
zamindar	landlord, receiver of rent
zarpeshgi	usufructuary mortgage
zamindari	permanently settled estate

Index

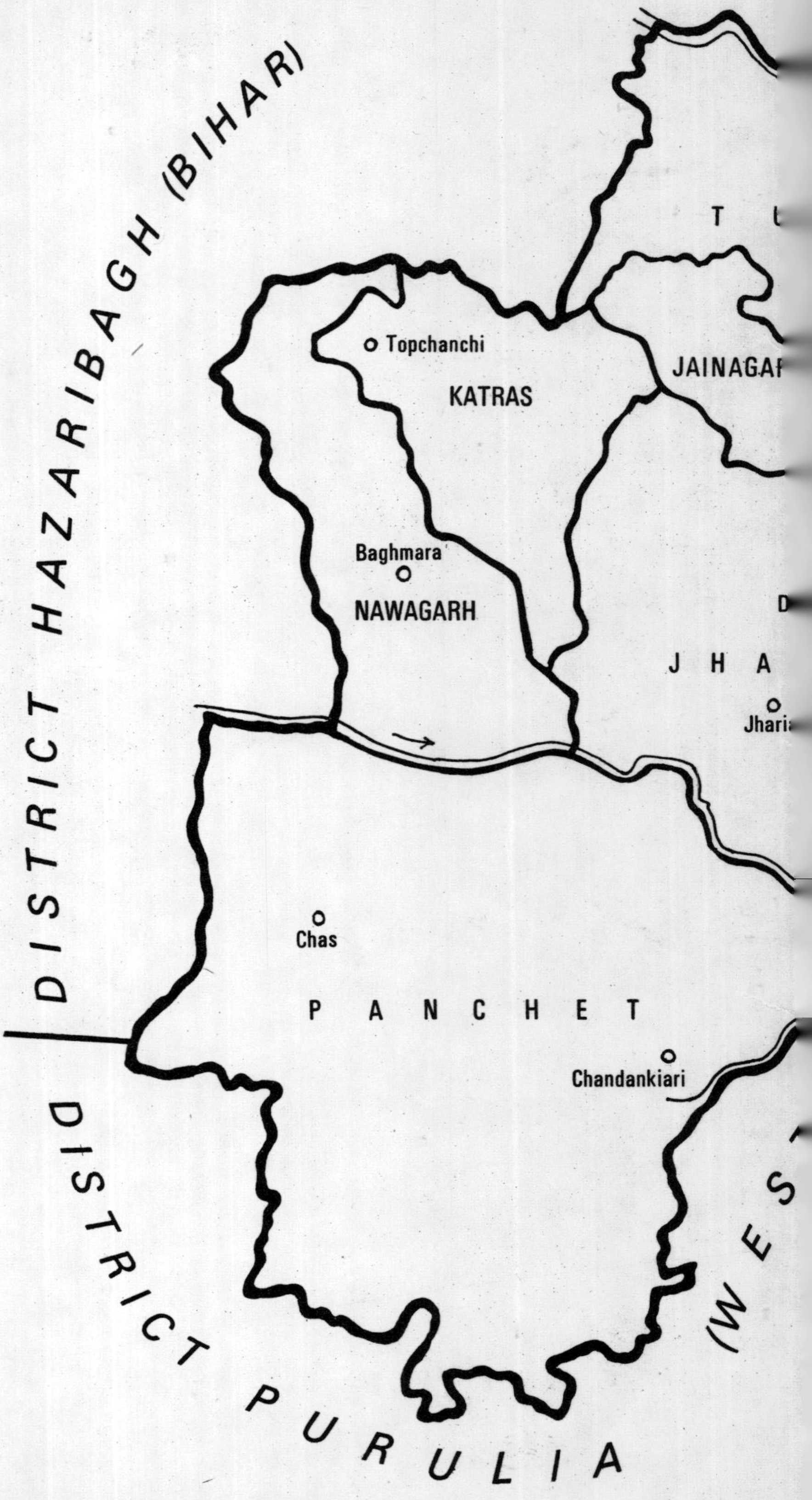
DISTRICT HAZARIBAGH (BIHAR)
Topchanchi
KATRAS
JAINAGAR
T U
Baghmara
NAWAGARH
D
J H A
Jharia
Chas
P A N C H E T
Chandankiari
DISTRICT PURULIA
(W E S